FOUNDATIONS FOR ADVANCING ANIMAL ECOLOGY

Wildlife Management and Conservation

Paul R. Krausman, Series Editor

Foundations for Advancing Animal Ecology

Michael L. Morrison
Leonard A. Brennan
Bruce G. Marcot
William M. Block
Kevin S. McKelvey

Published in Association with *THE WILDLIFE SOCIETY*

JOHNS HOPKINS UNIVERSITY PRESS | BALTIMORE

Johns Hopkins University Press
2715 North Charles Street
Baltimore, Maryland 21218-4363
www.press.jhu.edu

Library of Congress Cataloging-in-Publication Data

Names: Morrison, Michael L., author. | Brennan,
    Leonard A. (Leonard Alfred), author. | Marcot, Bruce G.,
    author. | Block, William M., author | McKelvey,
    Kevin S., author.
Title: Foundations for advancing animal ecology /
    Michael L. Morrison, Leonard A. Brennan, Bruce G.
    Marcot, William M. Block, Kevin S. McKelvey.
Description: Baltimore : Johns Hopkins University Press,
    2020. | Series: Wildlife management and conservation |
    Includes bibliographical references and index.
Identifiers: LCCN 2020003384 | ISBN 9781421439198
    (hardcover) | ISBN 9781421439204 (ebook)
Subjects: LCSH: Animal ecology—Research.
Classification: LCC QH541.2 .M68 2020 |
    DDC 591.7—dc23
LC record available at https://lccn.loc.gov/2020003384

A catalog record for this book is available from the British
Library.

Figures 2.4, 5.3, 5.4, 5.6, 5.7, 5.8, 6.1, 6.9, 7.4, and 7.6 were
created by VanDeWater Art and Design at the behest of the
authors.

*Special discounts are available for bulk purchases of this
book. For more information, please contact Special Sales
at specialsales@press.jhu.edu.*

Johns Hopkins University Press uses environmentally
friendly book materials, including recycled text paper that
is composed of at least 30 percent post-consumer waste,
whenever possible.

Contents

Preface *vii*

About the Authors *xi*

1 Operating Concepts for Animal Ecology 1

2 The Study of Habitat: A Historical
   and Philosophical Perspective 15

3 Heterogeneity and Disturbance 31

4 The Evolutionary Perspective: Linking Habitat
   to Population 70

5 Species Occurrence in Time and Space:
   Synthesis and Advancement 77

6 Managing Wild Animal Populations and
   Habitats in an Evolutionary and Ecosystem
   Context 111

7 Putting Concepts into Practice: Guidelines for
   Developing Study Plans 155

Index *181*

Preface

Our purpose in writing this book was lofty—namely, providing specific recommendations on how to substantially advance our field's approaches to studies of animal ecology. In other words, how do we maximize the probability that a species of wild animal will persist into the future? Such a goal clearly implies that, as authors, we collectively think that animal ecologists are failing to advance how we conduct research and apply that knowledge to successfully conserve wild animals. Animal ecologists are notorious for practices such as using vague and misleading terminology, taking the easy way out in designing and implementing studies, or failing to translate research findings into knowledge that natural resource managers can actually implement on the ground. It seems as though we believe that all we need is more money and additional time for more research, and everything will work out just fine. While these factors are important, what's of even greater consequence are clear definitions of terms, rigorous study designs, and controlled experiments wherever possible.

Yes, our opening paragraph here is harsh, but it is also valid. As we recount in detail, beginning in Chapter 1, this book focuses on individual animals and how they are organized as biological populations. That is the foundation. Building our studies around vague concepts such as "animal communities" clearly inhibits our understanding of animal ecology. A misunderstanding of broader temporal-spatial relation-

ships and differences between how humans and other animal species perceive their surroundings—the landscape—further inhibit how we approach research in animal ecology and make subsequent management recommendations. We are not the first to recognize the weaknesses in our approaches to animal ecology; we rely heavily on the writings of other scientists to make our case. Our book is, however, one of the few attempts to synthesize where we have been, where we are currently, and where we need to go with our studies of animals and their ecology.

We know that many readers will criticize us for being unrealistic, for expecting rapid and widespread changes in how we approach research design. A number of you will no doubt argue that it is difficult, and sometimes impossible, to quantify the biological population of wild animals for study. We welcome criticism and suggestions on how to do so; there are certainly cutting-edge genetic approaches, for example, that deserve more examination. Most likely there are better, or certainly additional, ways to measure behavioral interactions within and among biological populations beyond those recommended in this book. What we hope happens, though we have little confidence in it actually occurring, is that the majority of readers will agree with us and start including our recommended approaches in their ongoing and future studies. As a minimal—and first—step, we need to incorporate a discussion of the biological population(s) in all of our work and critically evaluate the utility of traditional terms and definitions we have become accustomed to accepting. Once we all engage in self-evaluation of our research habits, some very smart and clever people probably will devise ways around what has been described as the "boundary problem" in studies of animal ecology (see Chapter 1).

We purposefully omitted the term "wildlife" from the title of this book, because the popular connotation of the term is restricted to terrestrial vertebrates, whereas understanding the ecology of animals pertains to all taxa. In Chapter 1, we write in more detail about what constitutes wildlife and challenge our research and management communities to think broadly, as well as to look past traditional fields of zoological study and across ecosystem boundaries, rather than defending conventional study approaches.

We are confident that this book will be an invaluable resource to professionals and practitioners in natural resource management in the public and private sectors, including state and federal agencies, non-governmental organizations, restorationists, and environmental consultants. We say this because the chapters provide both a review of current practices and a demonstration of how we think those practices can be substantially improved. Our intended audience also includes upper division undergraduate and graduate students in courses and graduate seminars on animal ecology, wildlife ecology, wildlife management, land-use policy, and conservation biology.

The book focuses on concepts and theories in animal ecology, terminology, evolutionary underpinnings, and the history of wildlife-habitat studies. We begin by noting that classic studies of wildlife-habitat relationships were a reasonable and fruitful way to develop general understandings of where species occurred and the general environmental conditions there. They were almost always centered on habitat descriptions based largely on vegetation conditions. Regardless of the sophistication of the field methods and statistical analyses used, these classic studies were mostly time and place specific, but they did yield a wealth of valuable natural history information. We

then discuss the limitations of this approach and provide a rationale for a substantial refocus as we pursue additional knowledge on organisms and populations, as well as on how to conserve species of wild animals. We also provide a final chapter that circles back to our major theme—the biological population and boundary identification—as we provide guidance on how to place any study in the overall context of co-occurring animal populations and the environment. We offer this material as a road map for advancing how we study animals, which will substantially enhance the way in which we communicate among scientists and practitioners alike. A key here is communication between researchers and managers. Researchers must understand management objectives; managers must understand and embrace the information needed to move forward to meet those objectives.

In a forthcoming (2021) companion volume on applications, we will emphasize study design, including experimental approaches; where, when, and how to gather measurements for the study of animal ecology; measurements of behavior (behavioral ecology); modeling approaches; and an ending chapter on recommended future approaches in animal ecology. It also stands alone as a guide for practicing professionals as well as university students at all levels, if they have taken any general course in animal ecology. It is unique, however, in that it uses the concepts of the animal populations we emphasize in the present book.

Some of the topics we have included were previously developed in *Wildlife-Habitat Relationships: Concepts and Applications* (Morrison, Marcot, and Mannan 2006). There, we did touch on some of the themes that are expanded upon in this book, such as the need for clarity in terminology—most essentially, a clear working definition of habitat. Since that earlier book was written, however, greater exigencies for effective conservation have arisen. Science and management, now more than ever, need to find new approaches and partnerships, particularly in the face of accelerating climate change, large-scale disturbances, the degradation and fragmentation of natural areas, and a growing awareness of such major problems as increases in the illegal animal trade and the massive befoulment of entire ecosystems from discarded plastics. Our new effort in this book draws on additional years of thought, discussion, and our own field studies, driven largely by the realization that advancing animal ecology—and conservation itself—demands a fresh approach, a new perspective, and a return to the basics of understanding organisms in this rapidly changing global environment.

Numerous individuals provided support and guidance throughout this project. At Johns Hopkins University Press, we want to especially acknowledge senior acquisitions editor for life sciences, Tiffany Gasbarrini, for guiding this book. We also thank Esther Rodriguez, editorial assistant, for her substantial efforts in pulling the work together. Leonard Brennan would like to thank the C. C. Winn Endowed Chair in the Richard M. Kleberg Jr. Center for Quail Research, which provided the space and time for him to contribute to this project; as well as Joseph Buchanan, who provided review comments for several chapters. Bruce Marcot acknowledges support from the Pacific Northwest Research Station, USDA Forest Service. Kevin McKelvey acknowledges Jessie Golding, University of Montana, for reviews of several chapters. Joyce Van De Water provided graphic design and artwork throughout the book.

About the Authors

Michael L. Morrison, PhD
Professor and Caesar Kleberg Chair in Wildlife Ecology and
Conservation
Department of Rangeland, Wildlife, and Fisheries Management
Texas A&M University
College Station, Texas 77843-2258

Leonard A. Brennan, PhD
Professor and C. C. Winn Endowed Chair for Quail Research
Caesar Kleberg Wildlife Research Institute and Department
of Rangeland and Wildlife Sciences
Texas A&M University–Kingsville
Kingsville, Texas 78363

Bruce G. Marcot, PhD
Research Wildlife Biologist
USDA Forest Service, Pacific Northwest Research Station
620 SW Main Street, Suite 502
Portland, Oregon 97208

William M. Block, PhD
Emeritus Scientist
USDA Forest Service, Rocky Mountain Research Station
2500 S. Pine Knoll Drive
Flagstaff, Arizona 86001

Kevin S. McKelvey, PhD
Research Ecologist
USDA Forest Service, Rocky Mountain Research Station Forestry
Sciences Laboratory
800 E. Beckwith Avenue
Missoula, Montana 59801

FOUNDATIONS FOR ADVANCING ANIMAL ECOLOGY

1 — Operating Concepts for Animal Ecology

Let's be clear: the work of science has nothing whatever to do with consensus. Consensus is the business of politics. Science, on the contrary, requires only one investigator who happens to be right, which means that he or she has results that are verifiable by reference to the real world.

Michael Crichton, Caltech Michelin Lecture, January 17, 2003*

Morrison and Block (2021) reviewed what they called "the state of wildlife-habitat relationships" (see also Morrison 2012). They concluded that no major change has occurred in the way we approach studies of wildlife-habitat relationships over the past several decades, with advances coming primarily in the form of improved technology and statistical analyses. The basic template used in most studies of wildlife and habitat involves correlating some metric associated with animals (e.g., presence-absence, abundance) with measures of vegetation and other environmental features (e.g., soil depth, rock cover, presence of water). Morrison (2012) concluded that because most habitat studies collect additional examples of phenomena that were already well studied, the way we approach wildlife-habitat relationships has become an outmoded template for advancing management and conservation.

In a typical habitat study, a convenient study area is selected, vegetation and other environmental features are sampled, statistical analyses are applied, and the results are compared with studies done at different times and locations. Publication is then justified by extrapolating these findings to some unspec-

ified larger area. Then additional research needs are listed, along with a discussion of usually vague recommendations for management. In most cases, the animals observed are only a part of the biological population, and any inferences drawn from such data apply only to the place and time of that study (Hurlbert 1984). Additionally, a major impediment to advancing knowledge is that most of our data and the resulting inferences are being impacted in unknown ways by biotic and abiotic interactions that are either not included in the study or happening outside of the study area boundaries.

It thus seems more fruitful if we first determine the biological population to be examined and then sample characteristics of interest from that population, such as features of the environment and resources being used. We must also identify the relevant area (spatial extent) providing inputs of biotic and abiotic factors (e.g., nutrients, predators, competitors), which influence the distribution, abundance, and survivorship of populations. Identifying a biological population is difficult, because of discontinuities in distribution and individual animal movements. If we are to gather meaningful samples, however, we must understand how the animals are assorted into interacting groups of individuals. In addition to data on habitat, the focus has been on

*http://soilphysics.okstate.edu/quotes/

population parameters, such as abundance, density, or occupancy. Population demography and viability are rarely emphasized in wildlife-habitat studies. A better understanding of habitat associations and selections necessitates also studying population viability and demography, including reproduction and survival (Rodewald 2015).

We need to recognize that habitat can provide only limited insight into factors responsible for animal survival and fitness, and, thus, into population responses to changing environments. Within the broader niche concept, habitat is a positional component (i.e., where an animal is at a given time). Habitat studies are problematic, because the environmental features we measure at the location of an individual (or group of individuals) can stay the same, while the use of important resources by an animal within that location can change (Morrison 2009:64–65). For example, the position (location) of an individual can change between time periods, perhaps because of the arrival of a predator. In this situation, what is usually called "habitat" remains the same, but occupancy has changed, due to factors we did not measure (Dennis et al. 2003). An organism chooses to occupy a particular area or site in part because of the resources there. That is, whereas "habitat" refers to structural aspects of the environment. "resources" refers to trophic (food, water) and other components. In this way, resources at least partially define habitat. A species can deplete a patch of resources within an area, thus altering the habitat and, perhaps, occupancy there. This concept of habitat and resources is closely aligned with the concept of proximate and ultimate factors (sensu Hilden 1965), where a species occupies an area based on a set of cues (proximate factors), but ultimate factors (resource depletion, inter- and intraspecific competition, predation) influence its ability to survive and reproduce. A primary goal of this book, then, is to provide guidance on how we determine the physical boundaries within which we should conduct studies that will then provide reliable knowledge on factors driving animal persistence.

Everything obviously begins and ends with the individual (animal); that is the only entity that has a clear and unambiguous definition. Every other term we use is a concept that can get pretty fuzzy, including species and subspecies, population (especially its component parts, such as subpopulation), metapopulation, community, assemblage, guild, habitat, niche, landscape, and so forth. Broad concepts are easy to discuss but can be difficult to study effectively. We are not saying that these concepts should not be used, but we must be very careful in how we define and, especially, use these terms as a foundation for ecological and behavioral study. Below, we focus extensively on population.

What Is Wildlife?

The question "What is wildlife?" has recently been evolving along scientific, management, social, cultural, legal, and even ethical and esthetic dimensions. Traditional views of conservation and management in animal ecology have focused on wildlife solely in terms of terrestrial vertebrates: initially, game birds and mammals, then later adding non-game organisms of conservation concern, principally threatened and endangered species. Although much of wildlife management today still concentrates on these elements, since they remain important concerns for conservation action, the term "wildlife" is being broadened along several fronts.

For example, thanks to increasing interest in cumulative-effects analysis and the integration of evolutionary perspectives into wildlife science and animal ecology, the question "What is wildlife?" prompts one to ask which conditions of conservation interest, evolutionary significance, and ecological function should be included in an investigation (see Chapter 6). Should relatively scarce subspecies—particularly increasingly uncommon local endemics, such as Hawaiian coots, or 'alae ke 'oke'o (*Fulica americana alai*); Hawaiian gallinules, or 'alae 'ula (*Gallinula chloropus sandvicensis*); and black-crowned night-herons, or 'auku'u (*Nycticorax nycticorax ho-*

actli) of the Hawaiian Islands—be placed on an equal footing with full species for scientific, management, and legal priorities? Similar questions pertain in considering ecologically functional groups of organisms as subjects of both wildlife investigations and regulatory mandates.

"What is wildlife?" also raises important management concerns. Wildlife management—whether on private lands for hunted and nongame species, on state or federal public lands, on the lands of indigenous peoples, or on resource industry lands that include farms and commercial forests—has been mostly defined in terms of economic impacts from hunting or some other social factor, as well as the legal mandates related to protecting threatened and endangered species. Empirically, much of wildlife management centers on meeting regulatory edicts and exigencies related to potential or actual litigation. The focus of many wildlife management projects and policies is often driven by the lists of threatened, endangered, and rare or sensitive species, but they also often involve setting harvest regulations with dual goals: providing hunting opportunities, while also sustaining populations of the species being hunted. While other species (read non-hunted, or not listed as threatened or endangered) are not purposefully excluded from the realm of wildlife management, they often receive diminished or no formal management attention. Arguments have been raised that it is too complicated, and too expensive, to address the full list of wild animals in a particular area of conservation interest. Therefore, selected species of legal concern or consumptive interest should serve as signposts for the success or failure of conservation actions.

A number of other questions as to what qualifies as wildlife pertain to economic, cultural, ethical, and even esthetic dimensions. Economic issues center around questions of crop depredation, disease and pathogen organisms, forest insect pests, and similar direct assaults (or benefits) to natural or food resources and human health. Economic concerns focus on the indirect effects of providing habitat for threatened or endangered organisms, which can be seen to have adverse economic consequences for human communities and resource-use interests. Conservation and political issues revolving around spotted owls (*Strix occidentalis*) in the western United States are a classic example of a situation that has yet to be resolved. Cultural issues often concern the rights of indigenous peoples to use particular lands and habitats, and to hunt or gather plants and animals. Other cultural issues look at the traditional (although not necessarily indigenous) use of lands and wildlife populations for hunting purposes. Ethical issues can pertain to the general use of resources or to specific animal groups. Animal-rights interest groups generally focus on species that seem closer to humans in their degree of apparent sentience (including their expression of pain) and tend to leave out most fish, reptiles, amphibians, invertebrates, and plants. Artistic questions may deal with a sense of place, or with organisms of specific esthetic interest, and thereby exclude habitats and creatures not meeting such standards (e.g., some dangerous or venomous ones, such as snakes and spiders, as well as microorganisms).

It is clear that such a simple question as "What is wildlife?" has deceptively vast ramifications for guiding social priorities, legal decisions, directions for management, public education, and, ultimately, the future of organisms and habitats across the land. It is equally clear that no single definition of wildlife satisfies all such interests. We propose that the domain of wildlife ecology should expand to fully encompass all organisms collectively addressed by these dimensions in an ecosystem context (i.e., animal ecology, as noted in the title of this book; see also Cox and Merrill 2015; Grodsky et al. 2015). "Wildlife" should ultimately be defined as no less than the full array of living organisms of all taxonomic groups in terrestrial, riparian, and aquatic environments, including marine, estuarine, and below-ground ecosystems.

Biological Organization and Perception

Lindenmayer and Fischer (2006) cited biological organization and the human perspective as core issues to be clearly identified and defined in ecological work. "Biological organization" refers to whether the focus is on a single species or is an aggregate measure for several taxa, such as a multiple-species assemblage. Perspective is whether the focus is on the perception of the environment by non-human organisms or by humans (e.g., landscape, seen from the human or non-human animal perspective). They noted that miscommunication between scientists has been driven in large part by confusion over perspective and the level of biological organization in question.

In this book we take the approach that studies of animal ecology best begin with a focus on the behaviors and characteristics of individual organisms, which are then put into the context of how they may organize themselves into collections, such as with breeding pairs, leks, herds, flocks, populations, and so forth. This, then, provides the foundation for how studies on habitat translate into persistence of the species. Thus the foundation of this book is on how individuals can be viewed and studied in various contexts of collections of individuals—including addressing issues such as subspecies, subpopulations, and ecotypes—and on the spatial extent that should be considered. Once we know how a species of interest is distributed in space, we then know what to measure about them (i.e., the scales of measurements across appropriate dimensions of space, time, and resource gradients). Our application of concepts, including landscape and fragmentation, will build off of this foundation. We will take the perspective of the organism of interest in developing concepts and applications, while always keeping the potentially biased human perspective in mind. The human perspective is important, because if the ultimate utility of ecology is conservation, then we must be able to translate our work into actions that are effective and understandable to the public, policy makers, and legislators.

Lindenmayer and Fischer (2006) called habitat fragmentation an example of a "panchreston," which Merriam-Webster's dictionary (https://www.merriam-webster.com/) defines as "a broadly inclusive and oversimplified thesis that is intended to cover all possible variations within an area of concern." They diagrammed an approach to tackling the habitat fragmentation panchreston through a series of logically linked steps. They noted that their method could equally pertain to other areas of applied ecological research, including reintroduction biology and restoration ecology. These steps were framed as a sequence of logical questions.

- Which themes in the habitat fragmentation domain are to be examined?
- What processes and patterns within these themes are the key ones to investigate (e.g., habitat loss, habitat subdivision, or landscape connectivity)?
- What is the response being quantified: the presence or abundance of a particular species, the reproductive success of that species, or an aggregate measure, such as species richness?
- Which patterns and processes associated with landscape change might give rise to the patterns of abundance, breeding success, or assemblage composition that are being documented, and how can they be quantified?
- What is the appropriate conceptual landscape model that forms the backdrop against which particular hypotheses might be tested?
- Is the island model appropriate, or should the landscape be conceptualized as a series of species-specific habitat gradients?
- Which methods are appropriate for examining the issues most in need of testing?
- Given a particular research outcome, what might be the best management strategies to reverse the key negative problems stemming from landscape alteration?
- Is mitigation successful, and can monitoring provide useful new insights for future work?

We will approach each topic covered in this book in a stepwise manner similar to that outlined above by Lindenmayer and Fischer (2006), including identifying the level(s) of biological organization of interest and the perspectives involved in developing a conceptual framework that leads to a study plan.

The Importance of Terminology

Being able to communicate and pass information across generations is a hallmark of the human species. Accordingly, the Linnean system of nomenclature was developed to organize and communicate about organisms across different languages. Similarly, groups such as the American Ornithologists' Union (now the American Ornithological Society) developed the *Checklist of North American Birds* in an attempt to standardize avian common names (i.e., associate a scientific name with a particular common name). The topic of ecological terminology, which is central to communication, has been covered in the literature (e.g., Peters 1991; Wells and Richmond 1995; Hall et al. 1997; Morrison and Hall 2002; Hodges 2008). We will provide specific definitions of key terms, as well as literature support and a brief statement as to why we used that term. We discuss our rationale and approach below.

Hodges (2008) argued that recommendations to standardize ecological terminology or to develop rigid classificatory schemes in advance of our datasets do not further our science or its application. Her concern was that as knowledge is gained, we will naturally modify our definitions. She also argued that strict definitions fail to gain acceptance, because most scientists will not even be aware of suggested terminological revisions (i.e., not read the literature), readers will continually locate older references that could remind them of alternate definitions and uses, and different meanings will be applied to a concept because of variable contexts across studies. We certainly agree that these issues are an impediment to the standardization and use of terms. And it is clearly more important, regardless of the words used, that studies are rigorously designed and implemented, so as to produce reliable knowledge (Morrison et al. 2008). We also agree that definitions should be flexible and will,

naturally, change as knowledge of a concept is gained through study.

We disagree with Hodges (2008), however, in that polysemy and synonymy are not the major cause of confusion in papers in the ecological literature, nor are they impediments to ecological progress. Hodges (2008:41) concluded that "insightful language reviews will focus on developing more useful classifications of the concepts that our language imperfectly captures, rather than offering prescriptionist approaches to our terminology." We do not think Hodges (2008) was arguing against standardization per se, but rather against the acceptance of some inflexible definitional system that few would follow in any case. We fail to see the dichotomy between the definitions applied and the development of conceptual domains, because doing the latter will not cause more people to locate your paper or ignore previous uses! Regardless, we agree with Hodges (2008) that the focus should be on clarifying conceptual domains and developing a better classificatory ability.

Unfortunately, there remains a wide misuse of many key terms in ecology, which indicates a lack of scholarship by researchers and a dearth of rigor in the peer review process. As mentioned previously with regard to the term "habitat" (e.g., Hall et al. 1997; Morrison and Hall 2002; Guthery and Strickland 2015), authors of numerous papers mix terms with different meanings (e.g., "habitat" versus "habitat type"), misapply concepts (e.g., guild), consistently claim to study and advance knowledge of concepts they do not define and, most likely, do not even exist (e.g., community), and so forth. We think such failures are a symptom of why ecological knowledge does not seem to be advancing rapidly. As Hodges (2008) noted, we can usually determine, through reading a paper, what the author meant by the context of the study. Thus the problem might not be terminology per se, but rather a lack of clarity on what is actually being examined, as well as on what should be studied to address the stated goal (which usually includes the conservation of species).

Lindenmayer and Fischer (2006) stated that conservation biologists and ecologists need to be precise and consistent in their use of concepts and terms, in order to communicate effectively with each other and with colleagues from other disciplines. They highlighted the problem of miscommunication complicating environmental issues when there are public policy and legal implications. In the specific case of habitat fragmentation, they concluded that separating patterns and processes is crucial, in order to identify and quantify the underlying mechanisms threatening species. Most of the ecological literature abuses the term "fragmentation" and adduces it as a synonym for stressors, or even extinction; that is, fragmentation, however defined, is always bad. Rather, fragmentation can mean very different things once patterns and processes are identified, defined, and quantified. For example, fragmenting some otherwise contiguous environmental condition (habitat for a given species) by human activities indeed contributes to an extinction vortex (e.g., for northern spotted owls [*Strix occidentalis caurina*] and old-growth forest; or for undisturbed mature mixed forest and Siberian tigers [*Panthera tigris altaica*]). In contrast, fragmenting some environmental condition can help maintain the overall viability of a species or subspecies, such as avoiding synchrony of population cycles or the spread of pathogens.

Hodges (2008) noted that the operationalism of terms has a limit when applied to broad concepts at a high level of generality (see also Peters 1991). She concluded that formulating operational definitions for broad concepts is probably impossible, because such definitions are often so narrow that they are useful only in localized cases. We agree, and we think that a large part of the difficulty in advancing knowledge of the drivers of animal persistence comes from our propensity to focus on generalities that cannot guide meaningful research. Although generalities certainly contribute to developing broad concepts and theories, attempts to formulate research based on vague umbrella terms, without also including explicit definitions of the way they are being used, doom that research to irrelevancy.

Next, we list key concepts that will be prominent in this book, along with an explanation of our rationales for choosing them and for how they will be used. Additionally, throughout the text, when we reference literature using terms that do not follow our meaning for them, we will briefly note the difference, both to avoid confusion and to refrain from any extended justification of or debate about those terms. For example, if a paper referred to "guild," we will state that by our usage, they meant "assemblage." We will use the original intent—as best we can determine—of a term, and indicate when we deviate from such use, including why we have chosen to do so (e.g., see our comment below on "deme").

We do, however, begin with a lengthy—relative to other terms and concepts we cover—discussion of "population." This serves as a key foundation upon which the majority of this book will be developed. A critical point is differentiating groups defined by genetic criteria from those defined by ecological and demographic ones. For example, subspecies and populations are delineated primarily by genetic criteria; ecotypes, by ecological ones. It is generally accepted that subspecies are phylogenetically distinct groups, populations are intrabreeding groups with a limited gene flow to other populations, and ecotypes are conspecific groups with localized, similar ecological adaptations, regardless of their genealogical relationship. Other recognized clusters of individuals, such as bird flocks and mammal herds, are groups with common seasonal ranges and may or may not have a recognizable genetic differentiation. Each of these terms has utility, but it is important to define and use them consistently (Guthery and Strickland 2015). The level of differentiation needed to distinguish such groups will vary, but agreement on the type of data used (i.e., genetic versus ecological) is a necessary step toward more consistent classifications (Cronin et al. 2005).

Biological Population and Its Components

Cronin (2006) reviewed the complicated terminology that has been applied to intraspecific groups of animals, including the recent application of such word choices to regulatory (e.g., Endangered Species Act) applications. He concluded that the current use of intraspecies terms was often inconsistent and redundant, which led to much misuse and confusion regarding them in the literature. He recommended a simplification of the following commonly used words for application to wildlife studies and management (see Cronin 2006 for literature sources).

- Subspecies = Evolutionary Significant Unit (ESU) or Evolutionary Unit (EU).
- Population = ESU, EU, Management Unit (MU), or Distinct Population Segment (DPS).
- Subpopulation = local population, deme, ESU, EU, MU, or DPS.
- Metapopulation = population composed of subpopulations.

Herein we adopt Cronin's (2006) simplified terminology, which focuses on subspecies, populations, and subpopulations. He recognized, as do we, that there are problems inherent in the use of these words, yet the recent proliferation of terms has not led to an advancement in understanding or communication. This simplified categorization also adheres to the taxonomic practice of using the terms initially applied to a taxon; hence "subspecies," "population," and "subpopulation" have priority (i.e., their use preceded the others). Additionally, this practice is consistent in using the same prefix (sub-) for divisions of a higher category (i.e., species/subspecies, population/subpopulation). This usage also allows for a consideration of genetics, in addition to demography and geography, a concept we develop below. Another intraspecific category, ecotype—which does not require the identification of genetic ancestry—is also discussed below.

Regulations under the Endangered Species Act (ESA) explicitly refer to such entities as ESUs and DPSs. Under the ESA, an ESU is a population or group of populations that is substantially reproductively isolated from other conspecific populations and represents an important component of the evolutionary legacy of the species. Likewise, a DPS can be listed under the ESA. Although we are adopting Cronin's (2006) lexicon for ecological research, it is important to note that some terms still must be legally used and addressed.

An additional intraspecific classification is ecotype, which is defined as populations with convergent morphological, demographic, and behavioral adaptations to similar ecological conditions (see Cronin et al. 2005 for a review and examples of the material we summarize here). For instance, ecotypes have been identified in caribou (*Rangifer tarandus*) to include a small-bodied High Arctic form, a barren-ground tundra-dwelling form, a mountain form, and a forest-dwelling woodland form. Killer whales (*Orcinus orca*) are categorized into multiple ecotypes, with these groups differing in their morphology, behavior, and vocalizations (de Bruyn et al. 2013). The fact that ecotypic variation has been widely recognized in diverse groups calls for consideration of the topic here.

The critical distinction between ecotypes and subspecies is that the latter is defined by genealogical relationships. In other words, subspecies can also be ecotypes, but identifying an ecotype does not necessarily convey subspecific consideration. Thus ecotypes are conspecific groups with similar ecological adaptations to local conditions, regardless of their genealogical relationship. As noted by Cronin et al. (2005), opinions over the amount of differentiation needed to raise an ecotype to subspecies (or even species) status will vary among researchers. Our purpose here is to highlight the importance of recognizing that ecotypes do occur, and that different management strategies for their conservation, based on the geographic location of the population, will often need to be applied.

Linking Population and the Environment

As outlined at the beginning of this book, our primary purpose is to provide a framework for substantially advancing our understanding of wildlife. This framework must, therefore, focus on how to gain reliable knowledge about the factors and processes affecting the persistence of individual animals and their grouping into populations. As such, our first task is to explain the concept of biological population, because it is usually the single most important foundation we have for building our studies. We will continually refer to "the population" throughout this book.

Biologists studying ecology and evolution use the term "population" in many different ways. Common uses include a very vague definition, such as a collection of individuals belonging to a single species; a grouping identified only for a specific study and study area; and a gathering of animals that involves a geographic or administrative boundary, often adding a temporal restriction such as the breeding season for a breeding population (see Wells and Richmond 1995; Millstein 2010, 2014). As developed by Millstein (2009), if any arbitrary set of conspecifics can be a population, then the assumption is that selection and drift—processes that occur within biological populations—are also arbitrary. Populations are continually evolving in response to a host of interrelated biotic and abiotic factors; hence it is odd that scientists have generally treated this entity in such a cavalier manner. The substantial advancements made in the understanding of animal genetics over the past several decades are not being taken advantage of in most studies of animal ecology. For example, the term "deme" is fraught with confusing and often contradictory applications in the ecological literature, because its original intent has been modified over time (including even by such notable scientists as Mayr 1963). The current use of "deme," without a prefix, to mean a local interbreeding population is an incorrect use of the term (Briggs and Block 1981). We will not, therefore, use "deme," because of these issues.

Millstein (2014) reviewed and carefully considered many of the nuances of predominant ideas on the population concept, including especially those of Levins (1968) and Brandon (1990). She showed that the variability between the former and the latter in classifying environments as homogenous, coarse-grained heterogeneous, or fine-grained heterogeneous, for example, arose because they did not explicitly consider how to delineate the boundaries of a selective environment (i.e., the evolutionary perspective of how animals became adapted to a set of environmental conditions). She attempted, and we think largely succeeded, in blending aspects of various population concepts into a workable format that we adopt here. Independently from Millstein's (2014) work, and without explicitly discussing selection, Morrison (2012) also developed the case that wildlife management and conservation in general, and wildlife-habitat relationships more specifically, were failing to substantially advance knowledge and applications because of a broad failure to more explicitly address the biological population being studied.

It matters whether environments are viewed by organisms as being fine-grained or coarse-grained, because differences arise in the direction their selection will most likely take (see Millstein 2014:Table 1 for details and a literature review). Millstein (2014) showed that if boundaries are drawn in various ways, based on differing population concepts, you will probably make incorrect predictions, because the location of an environment's boundaries affects whether it is viewed as being fine-grained or coarse-grained (which will translate into how we gather measures of the environment). As we have argued above, to more fully understand what is driving species distribution and, ultimately, persistence we must move away (as much as possible) from using a human perspective of the environment and toward a view of how individuals and populations react to their surroundings. This led Millstein (2014) to argue for and develop a more principled way to draw the boundaries within which we evaluate fitness outcomes in relation to animals' occupancy of the environment. She explained that the term "selective environment" should not be taken to exclude other evolutionary processes, such as drift, from occurring in the same area.

Millstein (2014) suggested that, because a common foundation of population theory is that populations are a unit of evolution, they dictate the boundaries of the selective environment. That is, the boundaries of their environment are determined by the spatial location (extent) of the population (Millstein 2014:Fig. 1). Advantages of recognizing that populations bound their environments include the reasonable rationale that environments should be understood relative to the organisms that inhabit them. This also avoids the problem of using a multitude of often arbitrary (relative to the organism) boundary solutions, which would not be consistent with the different predictions in various types of environments. Lastly, it respects the types of population-level predictions of homogeneous, fine-grained, or coarse-grained environments, as well as acknowledging that these predictions reflect the common environmental conditions experienced by organisms of the same population (Millstein 2014).

Millstein (2014) continued to describe and define this population concept to delineate the environment in which species are evolving. She developed the Causal Interactionist Population Concept, or CIPC (see also Millstein 2010). CIPC characterizes populations in ecological and evolutionary contexts as consisting of at least 2 conspecific organisms that, over the course of a generation, are actually engaged in survival or reproductive interactions, or both. The boundaries of a population are the largest grouping for which the rates of interaction are much higher within that grouping than outside it. Both reproductive and survival interactions should be understood broadly. Relevant interactions include both unsuccessful and successful matings (interbreeding), offspring rearing, competition for limited resources, and cooperative activities. She noted that her qualification recognizes that other areas of study, such as statistics and biomedicine, may have alternative population concepts.

Thus organisms that are located within the same spatial area are part of the population if they are interacting with other conspecifics. A population is continuous over time if a later grouping of conspecifics is causally connected by survival or reproductive interactions to an earlier grouping. Individuals passing through a population, such as migrants or vagrants, but not causally interacting with them would not be part of that population (Millstein 2010). Properly establishing the boundaries of a population is central to accurately predicting its future trajectory; excluding interacting individuals (i.e., boundaries that are too small) or including non-interacting individuals (i.e., boundaries that are too large) would lead to misleading predictions (Millstein 2010).

Millstein (2014) continued to explain how CIPC was directly relevant to related population issues, such as the metapopulation. As reviewed by Hanski (1998), the foundation of the metapopulation concept follows Levins's (1969) description of metapopulation: a population of unstable local populations that inhabit discrete areas. In later chapters of this book (especially Chapters 3 and 4), we discuss the occurrence of metapopulations in nature. Millstein (2014) offered a scenario where there was local adaptation within 2 patches, with widespread interbreeding across patches. Then, in addition to reproductive interactions between individuals in the patches, there were significant processes (casual interactions) that enhanced survival rates within (but not between) patches. Since the rates of interactions enhancing survival within patches are significantly greater than the rates of causal interactions more generally, within CIPC there are 2 populations that may later become 2 subspecies or even species. Thus there are 2 homogeneous evolutionary environments whose boundaries are delineated by the spatial range of the causal interactions of the 2 populations. Since there are some interactions among the 2 populations, they form a metapopulation across a heterogeneous, probably coarse-grained environment. Therefore CIPC, unlike the population-as-deme view (see our comments on deme, above), preserved the idea that splitting a group into populations precedes splitting it into species. There is much debate on the manner in which speciation occurs, including allopatric, parapatric, peripatric, and sympatric means (Losos and Glor 2003). Our point here is to study how animals are distributed and interact without assuming a priori how individuals are distributed and divided into groups; our thesis does not require that a specific type of speciation has occurred or will happen. On the other hand, according to CIPC, if the rates of causal interactions within patches were not significantly greater than the rates of causal interactions among patches, then there would be only a single population evolving in a heterogeneous, probably fine-grained environment.

Thus metapopulations consist of at least 2 subpopulations ("local populations" in Millstein 2010) of the same species, linked by migration or dispersal, such that organisms occasionally change which subpopulation they are a part of. Rates of interaction within subpopulations are much higher than the rates of interaction among subpopulations (Millstein 2010). If such rates within local groupings of individuals are not significantly higher than those among local groupings, then they would represent a patchy population (not a metapopulation).

Ecosystem

The difficulty we have in designing plans for conservation (used here in a broad and general sense) is that we almost always must, for practical purposes, operate within physical boundaries designated by ourselves (i.e., humans). That is, we must respect public and private property boundaries. Additionally, there is always some limit to the funds available for implementing (and hopefully maintaining) management practices on the ground. Thus we are laying artificial boundaries over some—usually unknown and far too often not even considered—biological entity, such as a population or multiple populations.

In the text above, we have discussed the issue of considering the spatial extent of a biological population, including patchy populations and metapopulations, when we study animal ecology. In most cases, however, we are interested in multiple species, or certainly in considering the interactions among multiple species, when we design management plans. Additionally, we must consider the many biotic and abiotic factors that influence these species and, in turn, the physical boundaries within which we measure those factors. In other words, we must consider what is popularly considered "the ecosystem." Thus our next step is to determine how to bound it. Because an ecosystem is an ecological concept, it is difficult to strictly define and operationalize it. The boundaries of ecosystems are determined by the scientific questions being asked, and thus are not fixed by some inherent rule of nature. That said, we must establish some general rules for establishing ecosystem boundaries, so we can adequately replicate our studies and gain reliable knowledge over time.

Post et al. (2007) concluded that boundaries are set by discontinuities or steep gradients in the flux and flow of material and energy, as well as in interactions between populations of different species. Well-bounded systems are those such as lakes and islands, which makes delineating their boundaries straightforward, relative to more-open systems, including terrestrial habitats, estuaries, and streams.

Post et al. (2007) went on to discuss ways in which the boundaries of open systems can be delineated; we summarize their example here. They outlined a scenario where large inputs of nutrients were coming from the outside of an ecosystem boundary on short temporal scales, such as migrating animals moving through the system (Post et al. 2007:Fig. 2). Here, the system is larger than we had initially thought. In contrast, in cases where the system is highly productive and receives relatively little external input, such internal cycling determines the functional boundaries of the ecosystem. In open systems, each different question being asked will most likely necessitate the development of alternative definitions of ecosystem boundaries. Regardless, it is absolutely critical that we discuss these issues and the arrived-upon solution if we are to gain knowledge as we continue to study systems and apply the results to management.

Thus an ecosystem includes interactions or matter/energy flows from the larger area if and only if components are stronger or larger than those of the smaller area. Although developed in a slightly different context (i.e., the "land community" of Leopold 1949, which we have merged and adopted as a focus for the term "ecosystem" herein), this is the solution that Millstein (2018) adopted, because it takes into account most of the important causal processes that affects the future of the land under consideration. Millstein (2018) said "most of," because no biological system is truly closed, given enough time.

Habitat

Habitat is a species-specific concept and is not to be confused with land-cover patterns. As reviewed above, there are many historical reasons for this confusion, as well as the continuing misapplication of "habitat." Following our themes of biological organization and perspective developed above, it follows that researchers must specify whether the focus of their work is on land-cover patterns or on patterns of habitat for a particular individual species. For example, by definition, "habitat fragmentation" is a species-specific concept and not a land-cover one. Within the human-level perspective of a fragmented landscape, the habitat of certain species will be broken apart, while the habitat of other species will remain connected (e.g., think shrews versus elk, although resource selection and specificity also play key roles in determining the fragmentation and connectivity of species' habitats).

It is not our intent to examine and critique the use of all of the habitat terminology; various authors have thoroughly reviewed that topic (Hall et al. 1997; Morrison and Hall 2002; Kirk et al. 2018). Here, however, we do delve into the issues of microhabitat and macrohabitat, because of their pervasive use in the literature, which leads to our employment of the term herein. It is worth noting that in his review of microhabitat use in small mammal studies, Jorgensen (2004) concluded that the term was used differently among studies, which, he noted, introduced imprecision and confounded interpretation and comparisons. He also found that most studies were based on a small number of research plots, with a modest level of trapping effort. Jorgensen (2004) then relied on the terminology for macro- and microhabitat given by Morris (1987), who defined the former as the spatial area in which individuals perform all their biological functions, and the latter as being composed of environmental variables that affect individual behavior.

Prior to Jorgensen (2004), Hall et al. (1997) reviewed habitat terminology and concluded that macro- and microhabitat were relative and referred to the level (sensu Johnson 1980) at which a study is being conducted for the animal in question. Although they urged that macro- and microhabitat should be defined on study-specific and species-specific bases, they noted that macrohabitat was generally used to refer to broad-scale features, such as seral stages or zones of specific vegetation associations, which would usually equate to Johnson's (1980) first level ("order") of habitat selection. Microhabitat, then, usually referred to finer-scaled habitat features, such as would be important in levels 2–4 in Johnson's (1980) hierarchy. They concluded that it was appropriate to use micro- and macrohabitat in a relative sense, with the scales to which they apply stated explicitly.

Thus micro- and macrohabitat terminology is yet another example of researchers trying to fit what animals are doing along various continuums (e.g., niche axes) into discrete bins, without any direct relevance to the biological population(s) under study. Notably for our emphasis in this book, however, Jorgensen (2004:535) concluded that micro- and macrohabitat research had done "little to inform applied science and management needs that typically are more concerned with populations than individuals, and this review makes it clear that the relevance of the existing microhabitat research to population questions is very uncertain." Jorgensen (2004) clearly was aware of and concerned about the difficulty associated with studies of animals that were being conducted without any relevance to the biological population. Similar to our use of other terminology (e.g., "community") when referencing the literature, we cannot avoid using these micro- and macro- modifiers herein, but we do so fully understanding their limitations.

Landscape

As in our discussion above on habitat and habitat fragmentation, and our summary of the work by Lindenmayer and Fischer (2006), "landscapes" must be analyzed with a clear understanding of the biological entity (organizational issue) and perspective being considered. With clear operating definitions, we can determine how individual organisms (or assemblages of similar organisms) respond to variations (planned or not) in the structure of the environment. We are then able to discuss, say, how fragmentation of an area will impact individual organisms, rather than talking broadly and rather meaninglessly about "landscape fragmentation." Hodges (2008) noted that determining if a landscape was connected or fragmented depended on the species of interest and the attributes measured. Thus functional connectivity measurements address movement distances, patterns, and rates, whereas structural connectivity ones address attributes, such as forest density, that influence animal occurrence. What humans view as "not fragmented" on a large spatial extent (i.e., landscape scale) might actually be fragmented internally for any number of species.

As we define them here, landscapes are embedded within the broader ecosystem concept (see above). Landscapes represent a physical attribute of the ecosystem and are useful in describing how structural features of interest are arranged through space and time within an appropriately defined area.

As discussed by Morrison and Block (2021), designing habitat studies from the hierarchical perspective of how organisms view and use their environments is not equivalent to a landscape approach. Hierarchical habitat research is often designed by determining what conditions (i.e., "habitat") prevail in areas of increasing size around an animal (or group of animals). This hierarchy does place the location of an animal into a broader spatial context—for example, taking measurements at the nest substrate scale, at the territory scale, and at the watershed scale. This can be useful information, such as in predicting the potential distribution of the species across space. Yet these analyses are static, not dynamic, and are not reflective of population dynamics in space and time. This is because habitat descriptions per se do not capture anything about the various dynamic processes that interact directly and indirectly with a focal species or a group of species. See Morrison and Block (2021) for an additional development of this theme.

Synthesis

We approached the writing of this book with the goal of providing guidance on determining the factors influencing animal persistence over time. It is obvious that animal ecologists have learned a lot about where different species occur and have published innumerable descriptive papers on habitat and resource use in various places and times. Our underlying thesis, however, is these studies do not provide a framework for taking the next step in substantially advancing our understanding of animal ecology. We say this because they were not conducted in a manner that was relevant to the actual biological entities occurring in space and time; they represented a cookie cut from an unknown location in the dough.

In this chapter we have attempted to provide initial guidance on how we should determine the physical boundaries within which we should conduct studies. These will then provide reliable knowledge on factors driving animal persistence. We expand upon this guidance in the subsequent chapters. The studies needed to accomplish this goal will be substantially more difficult to conduct than a typical habitat study. They will usually require—scaled to the species of interest—sampling over larger spatial extents. They will often necessitate working across multiple landowners, not all of whom will allow access by researchers. Compromises will have to be made. Nevertheless, in the end we will start accumulating knowledge that allows us to start generalizing in a much more robust manner than currently possible.

This chapter sets the framework upon which we have built the remainder of this book. We have attempted to be as faithful as possible to the core definitions set forth in this chapter throughout the other chapters. We say "as possible," because the terminology accompanying many published studies is convoluted and inconsistent, which means that our interpretations might, at times, be off; we are not perfect! That said, the salient points from this chapter are that we must study how animals are distributed in space and time and not keep applying artificial boundaries to our studies. The biological population under consideration and its key associated constructs (e.g., subspecies, ecotype, metapopulation), as well as the associated physical area providing inputs to the population(s), must be considered if our studies are to inform science broadly. We must move away from study sites of convenience as much as possible. There is much work to be done.

LITERATURE CITED

Brandon, Robert N. 1990. Adaptation and environment. Princeton University Press, Princeton, NJ.

Briggs, D., and M. Block 1981. An investigation into the use of the "-deme" terminology. New Phytologist 89:729–735.

Cox, A., and E. Merrill. 2015. What taxa are appropriate for the Journal? Journal of Wildlife Management 79:527–528.

Cronin, M. A. 2006. A proposal to eliminate redundant terminology for intra-species groups. Wildlife Society Bulletin 34:237–241.

Cronin, M. A., M. D. MacNeil, and J. C. Patton. 2005. Variation in mitochondrial DNA and microsatellite DNA in caribou (*Rangifer tarandus*) in North America. Journal of Mammalogy 86:495–505.

de Bruyn, P. N., C. A. Tosh, and A. Terauds. 2013. Killer whale ecotypes: Is there a global model? Biological Reviews 88:62–80.

Dennis, R. L. H., T. G. Shreeve, and H. V. Dyck. 2003. Towards a functional resource-based concept for habitat: A butterfly biology viewpoint. Oikos 102:416–426.

Grodsky, S. M., R. B. Iglay, C. E. Sorensen, and C. E. Moorman. 2015. Should invertebrates receive greater inclusion in wildlife journals? Journal of Wildlife Management 79:529–536.

Guthery, F. S., and B. K. Strickland. 2015. Exploration and critique of habitat and habitat quality. Pp. 9–18 *in* M. L. Morrison and H. A. Mathewson, eds. Wildlife habitat and conservation: Concepts, challenges, and solutions. Johns Hopkins University Press, Baltimore.

Hall, L. S., P. R. Krausman, and M. L. Morrison. 1997. The habitat concept and a plea for standard terminology. Wildlife Society Bulletin 25:173–182.

Hanski, I. 1998. Metapopulation dynamics. Nature 396:41–49.

Hilden, O. 1965. Habitat selection in birds. Annales Zoologici Fennici 2:53–75.

Hodges, K. E. 2008. Defining the problem: Terminology and progress in ecology. Frontiers in Ecology and the Environment 6:35–42.

Hurlbert, S. H. 1984. Pseudoreplication and the design of ecological field experiments. Ecological Monographs 54:187–211.

Johnson, D. H. 1980. The comparison of usage and availability measurements for evaluating resource preference. Ecology 61:65–71.

Jorgensen, E. E. 2004. Small mammal use of microhabitat reviewed. Journal of Mammalogy 85:531–539.

Kirk, D. A., A. C. Park, A. C. Smith, B. J. Howes, B. K. Prouse, N. G. Kyssa, E. N. Fairhurst, and K. A. Prior. 2018. Our use, misuse, and abandonment of a concept: Whither habitat? Ecology and Evolution 8:4197–4208.

Leopold, A. 1949. The land ethic. Pp. 201–226 *in* A Sand County almanac and sketches here and there. Oxford University Press, New York.

Levins, R. 1968. Evolution in changing environments: Some theoretical explorations. Princeton University Press, Princeton, NJ.

Levins, R. 1969. Some demographic and genetic consequences of environmental heterogeneity for biological control. Bulletin of the Entomological Society of America 15:237–240.

Lindenmayer, D. B., and J. Fischer. 2006. Tackling the habitat fragmentation panchreston. Trends in Ecology and Evolution 22:127–132.

Losos, J. B., and R. E. Glor. 2003. Phylogenetic comparative methods and the geography of speciation. Trends in Ecology & Evolution 18:220–227.

Mayr, E. 1963. Animal species and evolution. Harvard University Press, Cambridge, MA.

Millstein, R. L. 2010. The concepts of population and metapopulation in evolutionary biology and ecology. Pp. 61–86 in M. Bell, D. Futuyma, W. Eanes, and J. Levinton, eds. Evolution since Darwin: The first 150 years. Sinauer Associates, Sunderland, MA.

Millstein, R. L. 2014. How the concept of "population" resolves concepts of "environment." Philosophy of Science 81:741–755.

Millstein, R. L. 2018. Is Aldo Leopold's "Land Community" an individual? Pp. 279–302 in O. Bueno, R. Chen, and M. B. Fagan, eds. Individuation, Process, and Scientific Practices. Oxford University Press, Oxford.

Morris, D. W. 1987. Ecological scale and habitat use. Ecology 68:362–369.

Morrison, M. L. 2009. Restoring wildlife: Ecological concepts and practical applications. Island Press, Washington, DC.

Morrison, M. L. 2012. The habitat sampling and analysis paradigm has limited value in animal conservation: A prequel. Journal of Wildlife Management 76:438–450.

Morrison, M. L., and W. M. Block. 2021. Wildlife-landscape relationships: A foundation for managing habitats on landscapes. In W. F. Porter, C. J. Parent, R. A. Stewart, and D. M. Williams, eds. Wildlife management and landscapes: Principles and applications. The Wildlife Society, Wildlife Management and Conservation. Johns Hopkins University Press, Baltimore.

Morrison, M. L., and L. S. Hall. 2002. Standard terminology: Toward a common language to advance ecological understanding and applications. Pp. 43–52 in J. M. Scott, P. J. Heglund, M. L. Morrison, J. B. Haufler, M. G. Raphael, W. A. Wall, and F. B. Samson, eds. Predicting species occurrences: Issues of scale and accuracy. Island Press, Washington, DC.

Morrison, M. L., W. M. Block, M. D. Strickland, B. A. Collier, and M. J. Peterson. Wildlife study design, 2nd edition. Springer-Verlag, New York.

Peters, R. H. 1991. A critique for ecology. Cambridge University Press, Cambridge.

Post, D. M., M. W. Doyle, J. L. Sabo, and J. C. Finlay. 2007. The problem of boundaries in defining ecosystems: A potential landmine for uniting geomorphology and ecology. Geomorphology 89:111–126.

Rodewald, A. D. 2015. Demographic consequences of habitat. Pp. 19–33 in M. L. Morrison and H. A. Mathewson, eds. Wildlife habitat and conservation: Concepts, challenges, and solutions. Johns Hopkins University Press, Baltimore.

Wells J. V., and M. E. Richmond. 1995. Populations, metapopulations, and species populations: What are they and who should care? Wildlife Society Bulletin 23:458–462.

2 — The Study of Habitat
A Historical and Philosophical Perspective

Eventually early scientists became aware of the accomplishments of their predecessors and immediate contemporaries. They were able to use each other's work, to build on it, to criticize it, improve on it, ignore it, sometimes parody it.

Hull (1988:75)

An animal's habitat is, in the most general sense, the place where it lives. Indeed, the word "habitat" is directly derived from the Latin verb *habitare*, meaning "to live, inhabit, or dwell." As a noun applied to wildlife, it literally means a place where wildlife live, or where they can be found, or where they create dwellings, such as dens or nests. All animals, except humans, can live in an area only if basic resources such as food, water, and cover are present, as well as if the animals have adapted in ways that allow them to cope with climatic extremes, competitors, predators, and other threats they encounter. Humans can live in areas even if these requirements are not met, because we can modify our surroundings to suit our needs or desires, and, more recently, because we potentially have access to resources such as food or building materials from all over the world. For these reasons, humans occupy nearly all terrestrial surfaces of the earth, but other species of animals are restricted to particular kinds of places.

The distribution of animal species among environments and the forces that cause these distributions have frequently been subjects of human interest, but for different reasons at various times. The primary purposes of this chapter are to review some of the reasons why people study the habitats of animals, and to outline how these reasons have changed over time. We also introduce the major concepts that will be addressed in this book.

Habitat Concepts of the Ancient World

The concept of habitat would almost certainly be familiar to a Pleistocene hunter. There were certain places and times when desirable target species would be present in numbers sufficient to hunt. A successful hunter would learn to recognize these areas and times and to exploit them (Fig. 2.1). Hunters would, by necessity, know animals well: the breeding grounds, migratory pathways, food preferences, and activity patterns for both their intended prey and for the other predators with which the humans competed. As such, they would be fully aware of the complex temporal aspects of habitat, such as multiyear patterns of abundance and scarcity, and seasonal shifts of resource availability and daily activity patterns—perhaps certain animals sleeping in shady areas during the heat of the day and foraging at dawn or dusk. If you were able to question these hunters about the habitat of an organism, they would probably ask you, "When?" Further, they would be well aware of the importance of juxtaposition. As with their own dwellings, they would recognize that

Figure 2.1   Pleistocene modern humans hunting a mammoth with spears. Pleistocene hunters would possess detailed knowledge about the biology of their prey. (*Hunting Woolly Mammoth*, https://cloudinary .com/. Creative Commons Attribution–ShareAlike 4.0 License, via Wikipedia.)

wildlife would choose areas with the proper combination of resources to facilitate their needs and temporal patterns.

While understanding habitat is essential for a hunter-gatherer, throughout recorded history humans, motivated by their curiosity, have observed and written or provided graphical representations about the habits of animals. Across the globe, hunter-gathers depicted wildlife on petroglyphs, alerting followers to the types of animals found in a particular location. Aristotle ([344 BC] 1862) observed animals and wrote about a wide variety of subjects, including breeding behavior, diets, migration, and hibernation. He also noted where animals lived and occasionally speculated about the reasons why.

> A number of fish also are found in sea-estuaries; such as the saupe, the gilthead, the red mullet, and, in point of fact, the greater part of the gregarious fishes. . . . Fish penetrate into the Euxine [estuary] for 2 reasons, and firstly for food. For the feeding is more

abundant and better in quality owing to the amount of fresh river-water that discharges into the sea. . . . Furthermore, fish penetrate into this sea for the purpose of breeding; for there are recesses there favorable for spawning, and the fresh and exceptionally sweet water has an invigorating effect on the spawn. (Aristotle [344 BC] 1862:124)

Interest in natural history waned after Aristotle's death. Politics and world conquest were the focus of attention during the growth of the Roman Empire; interest in religion and metaphysics suppressed creative observations of the natural world during the rise of Christendom (Beebe 1988). As a result, little new information was documented about animals and their habitats for nearly 1,700 years after the death of Aristotle. This is not to say that understandings of habitat were absent throughout this period. The stewards who were tasked to keep the king's forests full of deer and wild boar knew much about habitat—but they would have been largely illiterate. And people of all classes were certainly aware of habitat relationships. As Klopfer and Ganzhorn (1985) noted, painters in the medieval and Renaissance periods showed an appreciation for the association of specific animals with particular features of the environment. "Fanciful renderings aside, peacocks do not appear in drawings of moors nor moorhens in wheatfields" (Klopfer and Ganzhorn 1985:436). Similar appreciation is seen in the artwork from India, China, Japan (e.g., Sumi paintings), and elsewhere during this time. On the other hand, while keen observers noticed relationships between animals and their habitats during the Dark Ages, few of their insights were recorded in scientific or natural history formats.

The study of natural history emerged in the European Renaissance and was renewed in the seventeenth and eighteenth centuries. Most naturalists during this period, such as John Ray in the seventeenth century and Carl Linnaeus (Fig. 2.2) in the eighteenth century, were interested primarily in naming and classifying organisms in the natural world (Eiseley 1961). One of Linnaeus's chief accomplishments was the construction of relationship hierarchies above the species level. For example, he not only named humans as a unique species, *Homo sapiens*, but additionally classified us as primates (Koerner 1996). Linnaeus (1781) was also clearly interested in the dynamics of land and sea, as well as "the struggles of scientists in the Age of Enlightenment to reconcile their growing appreciation of the complexity of nature with biblical scripture" (Briggs and Humphries (2004:6). Pathfinders made numerous expeditions into unexplored or unmapped lands during this period, often with the intent of locating new trade routes or identifying new resources. Naturalists usually accompanied these expeditions, or sometimes they traveled on their own, collecting and recording information about the plants and animals they observed (Forster 1778). Many Europeans at this time also collected feathers, eggs, pelts, horns, and other parts of animals for "collection cabinets." Some cabinets were serious scientific efforts, but most were not, being intended largely as curiosities

Figure 2.2   Carl Linnaeus, also known after his ennoblement as Carl von Linné, was a Swedish botanist, physician, and zoologist who formalized binomial nomenclature (the modern system of naming organisms). He is known as the "father of modern taxonomy." His most important contribution was the creation of hierarchies above the species level. (*Carl von Linné*, Alexander Roslin, 1775, oil on canvas, Gripsholm Castle. Photo courtesy of the Nationalmuseum, Stockholm, Sweden, via Wikipedia.)

or status items used for displays. Nevertheless, new facts about the existence and distribution of animals worldwide were gathered during this period, and these advances in knowledge generated considerable serious curiosity about the natural world (Leclerc, Comte de Buffon, 1761).

During the nineteenth century, naturalists continued to describe the distribution of newly discovered plants and animals, but they also began to formulate ideas about how the natural world functions. Charles Darwin was among the most prominent among them. His observations on the distributions of similar species were a single set of facts, among many, that he marshaled to support his theory of evolution by natural selection (Darwin 1859). The work of

Darwin is highlighted here, not only because he recorded many new facts about animals, but also (and more importantly) because the theory of evolution by natural selection forms the framework of and foundation for the field of ecology.

The Birth of the Concept of Habitat

As noted above, the hunter-gatherer would likely have had a sophisticated and nuanced understanding of habitat. Unfortunately, in science, the word "habitat" has seldom been used in a manner that closely equates to the Pleistocene hunter's application of this concept. The study of wildlife habitat has largely changed our view of the hunter's understanding, from one which paid primary attention to the behavior of organisms and only secondarily used this behavioral knowledge to ascertain where an organism might be located, to one which primarily, frequently, and often exclusively determined where the organism was located and only weakly inferred what behaviors might be associated with the observed locations. In science, the study of habitat has always been, and is today, largely the study of where animals are. Behavioral research, or ethology, is mostly a separate branch of study. The types of behavior covered in this book are looked at in the context of some kind of habitat, such as foraging, nesting, and roosting. Splitting this integrated question of what is habitat into 2 distinct and disparate lines of research has, in turn, lead to uncertainty in the definition of habitat. Asking where an organism is, versus what an organism needs, has confused definitions of habitat for at least the last 100 years (Guthery and Strickland 2015).

The first uses of the word "habitat" that we can find in the scientific literature denote where a plant (Hillebrand 1888), animal (Woodhouse et al. 1852; Huxley and Martin 1889), or group of people (Dorsey 1885) could be found. These appeared as locations on maps (Dorsey 1885; Hillebrand 1888)—what we today might consider occurrences—or were denoted as generalized statements about the geographic area in which an organism could be found. Schaus (1899),

for example, in an article describing South American species of moths, under the category Habitat, stated in which country the organism could be found, such as "Habitat: Peru" (Schaus 1899:214). Similarly, both Carpenter (1914) and Woodhouse et al. (1852), when describing bird species in the United States, list habitats as being regions (e.g., "eastern North America" or "western Texas").

In the early twentieth century, however, the meaning of the term "habitat" began to broaden. Whereas prior common usage generally denoted an area where a particular species might be located, Oliver (1913) used the word to denote an area having certain characteristics, apart from any specific organism. For example, "narrow-mouthed salt marshes" and "broad-mouthed bays and mud-flats" are described as habitats. Clements (1916) championed the idea of species groups whose relationships changed over time in a definable manner, which he dubbed "succession."

At the same time, Priestley (1913) closely linked habitat to the presence of specific species. He suggested that habitat is a "rather intangible thing" (Priestly 1913:89) and suggested that an approach to making it tangible would be to locate it reliably and correlate its attributes to locations where the species was either found or not found. In modern practice, this would equate to a resource-selection approach (see below). Yapp (1922), noting this drift from a definition associated with *habitare* to one in which "habitat corresponds more or less closely to the English environment" (Yapp 1922:2), makes one of the earliest calls for formalism in its definition. More recently, Kirk et al. (2018) were still pleading for standardization in terminology. Over the course of Yapp's (1922) long and thoughtful article, however, it becomes clear that the definition had already morphed, to the extent that any single definition of habitat would necessarily be so general as to be largely vapid. Nevertheless, from this point forward, habitats would largely refer to the ecological conditions associated with an organism's presence, rather than a spot on a map showing where an animal occurred.

It is clear, when reading the earliest literature on habitat, that the primary goal was to categorize where an organism could be found and, secondarily, to infer causal reasons for its presence strictly from the patterns of its occurrence, rather than from a separate body of research. Like cladistics (and as we have noted in Chapter 1), habitat was seen as being hierarchal in nature. The broad spatial regions that filled these early writings as statements of "habitat" were simply high levels in the hierarchy. Yet in all cases, the focus was on where a species could be located. A species might be found in eastern North America, in eastern North American forests, in eastern North American pine forests, in eastern North American pine forests associated with sandy soils, and so forth. Any level in that hierarchy, from the highest to the lowest, could validly be called "habitat."

After the 1930s, habitat studies conformed more to the modern concept. Early work, such as that of Lack (1933), sought to specifically link animal communities to particular vegetation assemblages. By the 1940s, wildlife habitat was conceived of as vegetation composition, with or without the presence of wildlife. When discussing competition among bird species, Svärdson (1949:160) noted that *"Parus palustris and P. atricapillus* both breed in south Sweden, though in different habitats." He clearly separated the place ("south Sweden") from the definition of habitat (see Fig. 2.3 for an example of vegetation-based habitat). Thorpe (1945) posited that adaptive radiation can be achieved through habitat selection, even if geographic barriers are lacking—although, for his thesis to work, multiple habitats need to exist in the same area. In general, during the mid-twentieth century, habitats were largely undefined, other than being vegetation associations where organisms were found. Throughout this period, the group whose habitat definitions were most closely associated with the biological needs of the relevant organism were game managers. For example, Lay (1956), in discussing the effects of underburning in southern forests, described changes in habitat in terms of the effects on "desirable" and "undesirable" brush and grass spe-

Figure 2.3   By the 1940s, animal habitat was generally described by its vegetation composition, whether or not an organism was present. Complex, dense, mature forest in the southwestern United States, for example, is often described as Mexican spotted owl (*Strix occidentalis lucida*) habitat. We know that this image is of Mexican spotted owl habitat, as the photo contains a nest (*center*). (Photo courtesy of the US Department of Agriculture, Forest Service)

cies. For Lay (1956), habitat improvement meant giving white-tailed deer (*Odocoileus virginianus*) more to eat.

Development of the Habitat Selection Concept

In the early 1900s, curiosity about how animals interacted with their environment provided the impetus for numerous investigations into what are now called "ecological relationships." Interest in these relationships initially led to detailed descriptions of the distribution of animals, either along environmental gradients or among plant communities. Merriam (1890), for example, identified the changes that occurred in plant and animal species on an elevational gradient, and Adams (1908) studied changes in bird species that accompanied plant succession. Biologists living during this period postulated that climatic conditions and the availability of food and sites in which to breed were the primary factors determining the distributions of the animals they observed (see Grinnell 1917a).

Biologists in the early to mid-1900s, however, recognized that the distribution of some animals could not be explained solely on the basis of climate and essential resources. Lack (1933) was apparently the first to propose that some animals (in his case, birds) recognized features of appropriate environments, and that these features were the triggers that induced animals to select a place to live. Areas without these features, according to Lack (1933), generally would not be inhabited, even though they might contain all the necessary resources for survival. Lack's ideas gave birth to the concept of habitat selection and stimulated considerable research on animal-habitat relationships during the next 60 years.

Svärdson (1949) developed a general conceptual model of habitat selection, and Hilden (1965) later expressed similar ideas. Their models characterized habitat selection as a 2-stage process in which organisms first used general features of the landscape to select broadly from among different environments, and then they responded to subtler habitat characteristics to choose a specific place to live. Svärdson (1949) also suggested that factors other than those associated with the structure of an environment influenced selection. For example, whether an animal stays in or leaves a particular place could be influenced by conspecifics (Butler 1980), interspecific competitors (Werner and Hall 1979), and predators (Werner et al. 1983), as well as by features of the environment that are directly or indirectly related to the resources needed for survival and reproduction. Habitat selection, therefore, has come to be recognized as a complicated process involving several levels of discrimination and spatial scales, as well as a number of potentially interacting factors. The study of these factors and the behaviors involved in habitat selection has resulted in a wealth of information about why we find animals where we do. (See Stauffer 2002 for an overview of the history of habitat studies.)

The distribution of animals is also intimately tied to the concept of "niche." This concept has been defined in multiple ways over time (for historical overviews, see, e.g., Schoener 1989; Griesemer 1992; Pianka 1994) and continues to be the subject of much discussion. Grinnell (1917b) formally introduced the term when he was attempting to identify the reasons for the distribution of a single species of bird. His assessments included both spatial considerations (e.g., reasons for a close association with a type of vegetation), dietary dimensions, and constraints placed by the need to avoid predators (Schoener 1989). In this view, a niche included both positional and functional roles in the community. Elton (1927) later described a niche as the status of an animal in the community and focused on trophic position and diet. The views of niche articulated by Grinnell (1917b) and Elton (1927) often are contrasted, but Schoener (1989) argued that they had much in common, including the idea that a niche denotes a "place" in the community, dietary considerations, and predator-avoiding traits.

Hutchinson (1957) discussed the multivariate nature of the causes of animal distribution in his presentation of the n-dimensional niche. In this view, niche dimensions are represented by multiple environmental gradients. A given species (or population) can exist in only a subset of the conditions defined by all of the gradients (its potential or fundamental niche), but it may be further restricted in distribution (its realized niche) by predators and competitors. Odum (1959) viewed a niche as the position or status of an organism in an ecosystem resulting from its behavioral and morphological adaptations. His idea of a niche was dependent on both where an organism lived and what it did, but he separated, to some degree, habitat from niche with the analogy that an organism's "address" was its habitat, and its "profession" was its niche. More recent ideas about niches (e.g., MacArthur and Levins 1967; Levins 1968; Schoener 1974) considered niche axes as resources (i.e., those that are important for an animal), and niche as the combi-

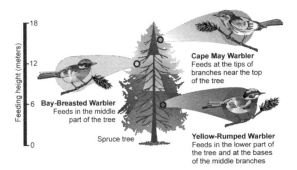

Figure 2.4 A representation of a bay-breasted warbler (*Setophaga castanea*), a Cape May warbler (*S. tigrina*), and a yellow-rumped warbler (*S. coronata coronata*), each using a different niche space (i.e., foraging axis). See MacArthur (1958) for details.

nation of several "utilization distributions" along those axes (Fig. 2.4).

The point of our brief review of the concept of niche is to illustrate that, although the term can be viewed in a variety of ways, most include elements that traditionally are considered to be a part of habitat. Thus studies designed to describe or define an animal's niche (of which there have been many) almost always elucidated animal-habitat relationships, as well.

It is reasonably easy to forget how limited scientists were when numerically evaluating data prior to the advent of computers. Calculating even simple descriptive statistics, such as arithmetic means, could take days of grinding tedium. Not surprisingly, the use of computers revolutionized the ways in which we evaluate patterns of animal occurrence. For example, resource selection is a fairly recent (Manly et al. 2002) formalism of an older idea associated with the habitat concepts associated with where an animal is. It is linked with a refinement of simple presence: habitat "quality"—generally undefined in terms of birthrates and survival that lead to a population increase or decrease (see Guthery and Strickland 2015)—is presumed to be greater with increases in organismal use. An area used more frequently is assumed to be better than one used less frequently. Often this idea of use is closely related to the idea of

"home range," loosely defined as an area of use across some particular time period. As such, resource-selection concepts grew out of a body of literature that defined patterns of use.

Early measures of home ranges included minimum convex polygons (Mohr 1947), which refer to the smallest convex polygon that encloses all locations of an animal where use could be demonstrated. This approach to bounding the range of the locations of an individual could be done without the need for fast computing speeds. Later, after computers revolutionized the statistical description of data, simple parametric alternatives emerged, such as bivariate normal distributions (Jennrich and Turner 1969). These approaches assumed that known occurrences were an independent and randomly drawn sample from a distribution. This assumption turned out to be problematic, however, due to the serial correlation associated with many occurrence datasets (Swihart and Slade 1985). Later, non-parametric approaches, such as kernel estimators (Worton 1989), were used. Kernel estimators assumed that each occurrence represented a local-use distribution, and that these local distributions could be added together to produce a general-use distribution. In essence, kernels are continuous forms, conceptually similar to histograms.

The statistical description of use, while helpful for delineating home ranges, was problematic for the task of assigning habitat quality. This difficulty led to the concept of describing use in comparison with available areas (Aebischer et al. 1993). An area having greater-than-expected use was assumed to represent better quality habitat. Often, more frequently used areas are referred to as being "preferred" (Mägi et al. 2009; Moore et al. 2018). Implementing the concept of use versus availability as a metric of habitat quality historically posed a number of statistical issues: the most common forms of occurrence data were spatially biased (e.g., occurrence databases primarily collected along roads), autocorrelated (e.g., closely spaced location data associated with GPS signals), or both. Some approaches (e.g., logistic regression) formally required data on presence and absence, when simple presence data was largely what was used (Keating and Cherry 2004).

There are 2 recent variants on the general theme of determining habitat quality based on the presence or densities of organisms. Occupancy modeling (MacKenzie et al. 2002, 2017) has been employed to define habitat use. This method draws on occurrence observations to produce a detection-corrected estimate of the likelihood of an animal being found within a particular space. Covariates can be included in the models, and habitat is inferred from the presence of particular covariates in highly ranked models. The use of occupancy models to describe habitats, while firmly in the modern tradition of capture-recapture sampling, is somewhat of a throwback to the earliest definitions of habitat, as the site where an organism could be found. Occupancy, at least in its original form, cannot be used to infer abundance (but see Royle and Nichols 2003; Royle et al. 2005).

The second variant on resource selection analysis uses spatial capture-recapture models (Efford and Fewster 2013; Royle et al. 2013a) to build animal density surfaces that can be related to habitats (Royle et al. 2013b). These models differ from standard resource selection approaches by assuming that the density of organisms in an area provides an index of habitat quality, rather than an assumption that the amount of time an individual spends in an area is indicative of habitat quality.

While a great deal of ink has been dedicated to controling and overcoming these issues over the years, the assumption that habitat quality can be gleaned from frequency of use or from density data has more fundamental and glaring problems, which have received less attention. The first difficulty (see Chapter 1) is that where you place the spatial boundaries for an analysis entirely determines the patterns of use and their statistical strength (Beyer et al. 2010). If you include large areas that are clearly not habitat and, therefore, are unused, you will produce strong patterns of "preference"; exclude these sites, and you may find no areas of preference. This understanding has led to the concept of hierarchal use versus available analyses (Johnson 1980), but, other than being a convention in the literature, there is no clear scientific rationale for specific boundaries. This is not a true hierarchy, but rather a series of arbitrary, nested spatial units.

Even more problematic is the assumption that habitat quality is closely associated with the amount of time an organism spends in an area, or with the density of organisms in a locale. The assumption is that use equates with quality, or even that use equates with preference. Some simple examples illustrate this difficulty. First, consider your home and assume that your location within the house is periodically tracked, producing location data that corresponds to the amount of time you spend in an area. This analysis will almost certainly find that your bedroom is the most preferred habitat. Your next choice is perhaps the room where you watch TV, or a home office. The bathrooms will be far less frequently utilized (i.e., less than expected). If you were to construct houses based on the usual understandings in resource selection analyses (as is common when planning species conservation), you would make houses with huge bedrooms and dens but probably no bathrooms. In a second example, let us say that you are studying a package delivery service and put GPS transmitters on top of the delivery trucks, in order to understand what they were up to. Your analysis would show the warehouse as being highly preferred, followed by the stop sign where all the trucks halt on their way to deliver packages, and then a couple of traffic lights along the major arterials that lead from the warehouse to the city. You would not, however, highlight the diffuse use associated with package delivery—which is the whole point of the enterprise. Here, the delivery trucks are similar in their patterns to central-place foragers (organisms, such as nesting birds, that have to foray out from a central site and return to it with acquired food); they do not "prefer" the stop lights, but their central-place activities force the use of these areas. While central-place foraging can be factored into the analyses

(Rosenberg and McKelvey 1999), this is based on external knowledge that the patterns observed are associated with central-place activity, such as providing food for young in a nest or den site. Lacking this formal understanding, one might seek to populate a city with stoplights, in order to provide critical resources that delivery trucks seem to "prefer."

These problems are more than theoretical. For example, Van Balen (1973), studying great tits (*Parus major*), found that their densities in deciduous woodlands were several times greater than those in coniferous forests. A study by Mägi et al. (2009), based on Van Balen (1973) and referring to deciduous woodlands as "preferred habitat," found that reproductive success in those woodlands was extremely low. Van Horne (1983) made this case more generally: assessments of habitat quality should be primarily based on the demographic characteristics of a population, rather than its density. If we applied the density criterion to human habitats, we would find that the "preferred" locales existed in mega-city slums. Similarly, in the animal world, there may be instances where "sinks" may harbor greater densities of individuals than do "sources," and thus be conflated with higher habitat quality (Van Horne 1983).

It is important to remember that this entire trajectory of approaches, from simple statements of habitats as countries where the organism is found, to spatial capture-recapture models using derived density surfaces to infer where organisms concentrate, are all simply variants of habitats equating to where organisms exist. While papers measuring habitat selection almost always provide ideas about the biological and demographic underpinnings of the observed occurrence patterns, these suppositions are not formally supported by selection analyses (Garshelis 2000). The presence of a cavity-nesting bird in older forests may be associated with the presence of cavities, or it may have an alternative explanation. Because demography is inferred, rather than directly measured, in resource selection analyses, inferences to organismal biology are limited. This happens despite the fact that resource selection represents a large portion of the total literature in wildlife biology and is supported by an immense and extremely sophisticated body of both observational technologies and statistical methods. As a result, it is not too surprising that modern, sophisticated analyses of use have not produced many novel conclusions concerning habitat relationships (Wiens 2016).

The long history of casual use of the word "habitat" is associated with a desire to infer more than the location of organisms within a species. Given this history, and the failure of newer methods to confront the fundamental limits associated with knowing only where an organism was, it is not surprising that use of the term "habitat" has continued to be haphazard. For example, Hall et al. (1997; see also Morrison and Hall 2002; Kirk et al. 2018) found that in describing habitat, most authors did not clearly define what they meant by that term, and many did not use the word consistently within the same paper. We consider the expectations of Hall et al. (1997) to have been modest. They defined habitat as the "resources and conditions in an area that produce occupancy"; habitat use as "the way an animal uses (or 'consumes,' in a generic sense) a collection of physical and biological components"; and habitat selection as "a hierarchical *process* involving a series of innate and learned behavioral decisions made by an animal about what habitat it would use at different scales of the environment" (Hall et al. 1997:175).

These definitions from Hall et al. (1997) separate "selection" from "preference," which is the time integral of a selection process. Note that preference is what is almost uniformly measured. From this, selection is inferred. Additionally, Hall et al. (1997: 175) defined habitat availability as the "accessibility and procurability of physical and biological components of a habitat." While these definitions are clear, other than "selection" and "preference," which are almost never actually measured, they still fall far short of any real relationship with the behaviors that ultimately link to the demographics of an organism (Block and Brennan 1993). It is a bit alarming that most of the papers reviewed by Hall et al. (1997)

failed to meet even these relatively low and achievable standards.

While it may seem odd to be describing a thing without clearly defining what that thing is, we have found that this practice is common in biology. Wade et al. (2015), for example, found that most papers discussing habitat connectivity failed to describe what sort of connectivity they meant. Just as there are many functional relationships associated with habitat, there are numerous functional needs for connectivity. Connectivity can be a means to obtain daily resources, facilitate seasonal movements to locate shifting resources, allow metapopulation demographic rescue, maintain a long-term exchange of genetic material, or to serve other functions. In this book, we will repeatedly confront the fact that most of the basic biological concepts of ecosystem, population, habitat, connectivity, and the like are seldom operationally defined within studies. We assert that failing to do so hides a plethora of sins of omission and fundamentally hampers the forward progress of ecology, particularly in general and animal ecology.

Regarding the definition of habitat, technological advances over time have led to new methods of observing animal locations, and expanding computer capabilities have provided furthered statistical analyses of the resulting patterns of occurrence. While these new technologies have allowed animal occurrences to be evaluated—by orders of magnitude— more rigorously than they had been in the past, the fundamental thinking behind habitats remained remarkably similar. What perhaps changed was a clear understanding of the limits to biological inferences associated with knowing where an organism could be found.

Societal Utilization of Habitat Concepts

Habitat concepts affect the business of society mainly when humans seek to exploit wildland resources and increase the abundance of target organisms. The earliest humans relied, in part, on killing animals for survival. They recognized and exploited the patterns of association between the animals they hunted and the kinds of places where these animals were most abundant. They also manipulated habitats to improve hunting. The use of fire by Native Americans altered the ecosystems in which they lived (Botkin 1990) and influenced the number of animals they hunted. Similarly, people who later made a living by trapping and hunting, or who could afford the luxury of hunting for sport, knew where to find animals and sought to enhance those environmental features that influenced the abundance of game species. For example, Marco Polo reported that in the Mongol Empire in Asia, Kublai Khan (AD 1215–1294), increased the number of quail and partridges available to him for falconry by planting patches of their foods, distributing grain during the winter, and controlling cover (Leopold 1933).

Not until people began to attempt applying biology systematically to the management of game as a "crop" in the early 1900s did they realize that "science had accumulated more knowledge of how to distinguish one species from another than of the habits, requirements, and inter-relationships of living population" (Leopold 1933:20). An absence of information about the habitat requirements of most animals and a desire to increase game populations by manipulating the environment stimulated detailed investigations of the habitats and life histories of game species. Stoddard's (1931) landmark book on northern bobwhites (*Colinus virginianus*), and work on pheasants (Errington and Hammerstrom 1937), exemplify early efforts of this kind. Life-history studies similar to Stoddard's (1931) have been conducted on most game animals in North America between 1930 and today (e.g. Bellrose 1976; Wallmo 1981; Thomas and Toweill 1982), but many of these publications only summarize general habitat associations and do not identify core habitat components.

While early uses of habitat concepts were applied almost exclusively to harvested species, the vast expansion of humankind during the twentieth century led to unprecedented changes in the conditions and amounts of wildland worldwide. Societal alarm at the rapid changes in landscape conditions and the precipitous declines of many species led to a series of laws designed to protect species and the habitats that support them (Leopold 1933). The following summary pertains to US history; it is beyond the scope of our book to review public interest and environmental law in other countries, but similar legislation has been passed by many nations.

Public interest early in the twentieth century focused on game animals, and some laws passed in the 1930s reflected this. The Migratory Bird Hunting Stamp Act and the Pittman-Robertson Federal Aid in Wildlife Restoration Act, for example, primarily taxed sportsmen and provided funds for the management of waterfowl and other hunted species. As noted in the previous section, information needed for the management of these species stimulated efforts to describe and quantify their habitats.

An increase in environmental awareness during the 1960s and 1970s broadened the scope of the kinds of animals the general public was concerned about. Animal species not hunted for sport and without any other apparent economic utility also started being perceived as having value. (The ethical rationales underlying these values are discussed in the next section.) Among the laws passed during this period were the National Environmental Policy Act, the Endangered Species Conservation Act, the Federal Land Policy and Management Act, and the National Forest Management Act (Bean 1977). Legislators designed these laws, in part, to ensure that all wildlife species and other natural resources were considered in the planning and execution of human activities on public lands.

Knowledge about the habitats of animal species obviously is required before the effects of an environmental disturbance can be fully evaluated, before a refuge for an endangered species can be designed, or before animal habitats can be maintained on lands managed under a multiple-use philosophy. Biologists responded to the need for information about habitat requirements by studying, often for the first time, numerous species of non-game (not hunted) animals

and by developing models to help predict the effects of environmental changes on animal populations (e.g., Verner et al. 1986). The Endangered Species Act added to the lexicon by coining the term "critical habitat" for "physical or biological features (I) essential to the conservation of the species and (II) which may require special management considerations or protection" (Endangered Species Act Amendments 1983:2).

In a democracy, laws are the clearest manifestation of general societal goals. For this reason, the passage of environmental laws has led to broad changes in land management that, in turn, have fostered increasing abundances of many species (Flather et al. 2000). It should be noted, however, that laws require formal rules: one needs to know unambiguously whether a law has been broken. Such laws tend to define goals, particularly habitat goals, in easy-to-measure (and generally crude) metrics, such as hectares of habitat created or retained, and kilometers of stream "restored." These needs, in turn, lead to dichotomous designations, such as where an area is either considered "critical habitat" or is not, with endless haggling over the obviously arbitrary boundaries between what is to be conserved and what is not. While so much effort and money have been expended to provide backing for these legal decisions, these actions have probably not had a beneficial effect on the study of habitat.

Public interest in the non-consumptive use of animals has not waned in recent years. In the United States in 2001, 66 million people (over 16 years of age) spent more than $38.4 billion observing, feeding, or photographing wildlife (USFWS 2003). The funding mechanisms for managing animals in the United States, however, have not kept pace with the broadening umbrella of public interest (Jacobson et al. 2007). Many state fish and game agencies have developed non-game management programs that emphasize identifying and managing habitats, but these programs often are limited by inadequate funding, and the sources of such monies are, with rare exception, not broad based or user related.

The Ethics of Habitat Conservation

Another impetus for studying habitat partly underlies the public interest and environmental laws outlined in the previous section and relates to an ethical concern for the future of wildlife and natural communities (Schmidtz and Willott 2002). This concern is, in part, a humanistic one, insofar as the health of natural systems affects our use and enjoyment of natural resources in the broadest sense. From a utilitarian viewpoint, the globe is our habitat, and its health, as well as that of all within it, directly relates to our own (Fig. 2.5). Ethical concerns, however, transcend humanism, in that wildlife and natural communities are intrinsic to the world in which we have evolved and now live, as Leopold (1949) sketched out in his concepts of a land ethic. Writers of legal as well as ethical literature have argued that nonhuman species have, in some sense, their own natural right to exist and grow (e.g., Stone 1974, 1987). The study of wildlife species and their habitats in this context may deepen our appreciation for an ethical responsibility to other species and natural systems.

Why should we be concerned about species and habitats that offer no immediate economic or recreational benefits? Several philosophical arguments provide complementary and sometimes conflicting rationales. One viewpoint argues for conserving species and their environments because we may someday learn how to exploit them for medical or other benefits (future option values). Another angle argues for preserving species for the unknown (and unknowable) interests of future generations, but we cannot speak for the desires of our not-yet-born progeny who will inherit the results of our management decisions.

Generally, a traditional conflict has pitted ethical humanism against humane moralism. Ethical humanism, as championed by W. K. C. Guthrie, Immanuel Kant, John Locke, Thomas More, and Thomas Aquinas, argues that animals are not "worthy" of equal consideration; they are not "up to" human levels, in that they do not share self-consciousness and personal interests. In effect, this argument allows us to subjugate wildlife and their habitats. Kant argues as such. He advanced his idea on a so-called deontological theme (from the Latin *deos*, meaning "duty"). That is, rights—specifically human rights—allow us to view animals as having less value, because they are less rational (or are arational). Thus we humans have a duty to manage species and the freedom to subjugate them. Indeed, conflicts exist between human values and wildlife (Messmer 2009). Wildlife populations that compromise agricultural production, or those considered to be pests or to transmit diseases,

Figure 2.5 Humans and wildlife are increasingly forced to occur in close proximity and share certain resources. Here, visitors interact with Rocky Mountain elk (*Cervus canadensis nelsoni*) in Grand Canyon National Park. (Photo courtesy of the National Park Service)

may require special attention (Messmer 2009; Decker et al. 2012). To address these conflicts, wildlife management may need to shift from sustaining or increasing populations to also mitigating conflicts (Messmer 2009).

On the other hand, humane moralism, as championed in part by Stone (1974, 1987), Regan (1983), and Singer (1990), argues that animals deserve the focus of ethical consideration. According to this argument, humans are moral agents. Animals and, by extension, their habitats require consideration equal to that given to humans, even if they do not ultimately receive equal treatment.

There also is a third ethical stance, one that may serve as an impetus for studying and conserving wildlife and their habitats: an ecological ethic. This ethic, as proposed by Fladder and Callicott (1992), was most eloquently advanced in *A Sand County Almanac* (Leopold 1949), although elements of Leopold's philosophy (and much fuller philosophical expositions) can be traced to Henri Berson, Teilhard de Chardin, and John Dewey. In this view, the focus of ethical consideration is on both the individual organism and the community in which it resides. Concern for the community is the essence of Leopold's (1949) ecological ethic, a holistic one that focuses on the relationships of animals with each other and with their environment. He wrote of soil, water, plants, animals, oceans, and mountains, calling each a natural entity. In his view, animals' functional roles in the community—not solely their utility for humans—provided a measure of their value. By extension, then, in order to act morally, we must maintain our individual human integrity, our social integrity, and the integrity of the biotic community.

Following such an ecological ethic, a concern for the present and future conditions of wildlife and their habitats motivated us in writing this book. The sad history of massive resource depletions, including extinctions of plant and animal species and the large-scale alteration of terrestrial and aquatic environments, must, in our view, strengthen a commitment to further out understanding of wildlife and their habitats. Understanding is the necessary prelude to truly living by an ecological ethic.

Changes in environmental conditions on earth, advances in technological devices and analytic methods, and the potential need for new philosophical and conceptual approaches in research and management require wildlife biologists do their best to keep abreast of new ideas. We suggest that the scale on which we conduct research and management be examined rigorously, and potential ways be explored, in order that the concept of niche may help focus research in the future. We also call for a more complete integration of the overlapping fields of wildlife ecology and management, conservation biology, and restoration ecology. Changes in the way we educate students and professionals must precede integration of this kind (see Chapter 6).

Society's views of the world and how it works are likely to shift over time, as current explanations are replaced by better ones. We were motivated to write this book because we concluded that the study of animal ecology was not advancing quickly enough to keep pace with our rapidly changing world-wide environment. We hope that, no matter what alterations occur, our readers—current or future conservationists in the broadest sense—remain tied to an ecological land ethic and continue to pursue and provide vital, productive habitats for wildlife and humans alike.

LITERATURE CITED

Adams, C. C. 1908. The ecological succession of birds. Auk 25:109–153.

Aebischer, N. J., P. A. Robertson, and R. E. Kenward. 1993. Compositional analysis of habitat use from animal radio-tracking data. Ecology 74(5):1313–1325.

Aristotle. [344 BC] 1862. Historia animalium. H. G. Bohn, London.

Bean, M. J. 1977. The evolution of national wildlife law: Report to the Council on Environmental Quality. Stock No. 041-011-00033-5. US Government Printing Office, Washington, DC.

Beebe, W., ed. 1988. The book of naturalists. Princeton University Press, Princeton, NJ.

Bellrose, F. C. 1976. Ducks, geese, and swans of North America. Stackpole, Harrisburg, PA.

Beyer, H. L., D. T. Haydon, J. M. Morales, J. L. Frair, M. Hebblewhite, M. Mitchell, and J. Matthiopoulos. 2010. The interpretation of habitat preference metrics under use-availability designs. Philosophical Transactions of the Royal Society, B: Biological Sciences 365:2245–2254.

Block, W. M., and L. A. Brennan. 1993. The habitat concept in ornithology: Theory and applications. Current Ornithology 11:35–91.

Botkin, D. B. 1990. Discordant harmonies: A new ecology for the twenty-first century. Oxford University Press, New York.

Briggs, J. C., and C. J. Humphries. 2004. Early classics. Pp. 5–13 in M. V. Lomolino, D. F. Sax, and J. H. Brown, eds. Foundations of biogeography. University of Chicago Press, Chicago.

Butler, R. G. 1980. Population size, social behavior, and dispersal in house mice: A quantitative investigation. Animal Behavior 28(1):78–85.

Carpenter, V. 1914. Some common birds. American Midland Naturalist 3(10):298–304.

Clements, F. E. 1916. Plant succession: An analysis of the development of vegetation. Publication No. 242. Carnegie Institution of Washington, Washington, DC.

Darwin, C. 1859. The origin of species. Penguin, New York.

Decker, D. J., W. F. Siemer, D. T. Evenson, R. C. Stedman, K. A. McComas, M. A. Wild, K. T. Castle, and K. M. Leong. 2012. Public perceptions of wildlife-associated disease: Risk communication matters. Human-Wildlife Interactions 6:112–122.

Dorsey, J. O., 1885. Omaha sociology. Third Annual Report. US Bureau of American Ethnology, Washington, DC.

Efford, M. G., and R. M. Fewster. 2013. Estimating population size by spatially explicit capture-recapture. Oikos 122:918–928.

Eiseley, L. 1961. Darwin's century. Doubleday Anchor, Garden City, NY.

Elton, C. 1927. Animal ecology. Sidgwick & Jackson, London.

Endangered Species Act Amendments of 1982. 1983. Public Law 97–34, 96 Stat. 1411. US Government Printing Office, Washington, DC.

Errington, P. L., and F. N. Hamerstrom. 1937. The evaluation of nesting losses and juvenile mortality of the ring-necked pheasant. Journal of Wildlife Management 1:3–20.

Fladder, S. L., and J. B. Callicott, eds. 1992. The river of the Mother of God and other essays by Aldo Leopold. University of Wisconsin Press, Madison.

Flather, C. H., S. J. Brady, and M. S. Knowles. 2000. Wildlife resource trends in the United States: A technical document supporting the 2000 RPA assessment. General Technical Report RMRS-GTR-33. US Department of Agriculture, Forest Service, Rocky Mountain Research Station, Fort Collins, CO.

Forster, J. R. 1778. Observations made during a voyage round the world, on physical geography, natural history, and ethic philosophy. 1966 reprint by N. Thomas, H. Guest, and M. Dettelbach, eds. University of Hawai'i Press, Honolulu.

Garshelis, D. L. 2000. Delusions in habitat evaluation: measuring use, selection, and importance. Pp. 111–164 in L. Boitani and T. K. Fuller, eds. Research techniques in animal ecology: Controversies and consequences. Columbia University Press, New York.

Griesemer, J. B. 1992. Niche: Historical perspectives. Pp. 231–240 in E. Fox-Keller and E. A. Lloyd, eds. Keywords in evolutionary biology. Harvard University Press, Cambridge, MA.

Grinnell, J. 1917a. Field tests of theories concerning distributional control. American Naturalist 51:115–128.

Grinnell, J. 1917b. The niche relations of the California thrasher. Auk 34:427–33.

Guthery, F. S., and B. K. Strickland. 2015. Exploration and critique of habitat and habitat quality. Pp. 9–18 in M. L. Morrison and H. A. Mathewson. eds. Wildlife habitat conservation: Concepts, challenges, and solutions. John Hopkins University Press, Baltimore.

Hall, L. S., P. R. Krausman, and M. L. Morrison. 1997. The habitat concept and a plea for standard terminology. Wildlife Society Bulletin, 26:173–182.

Hilden, O. 1965. Habitat selection in birds. Annales Zoologici Fennici 2:53–75.

Hillebrand, W. 1888. Flora of the Hawaiian Islands. Williams & Norgate, London.

Hull, D. L. 1988. Science as a process: An evolutionary account of the social and conceptual development of science. University of Chicago Press, Chicago.

Hutchinson, G. E. 1957. Concluding remarks. Cold Spring Harbor Symposium on Quantitative Biology 22: 415–427.

Huxley, T. H., and H. N. Martin. 1889. A course of elementary instruction in practical biology. Macmillan, New York.

Jacobson, C. A., D. J. Decker, and L. Carpenter. 2007. Securing alternative funding for wildlife management: Insights from agency leaders. Journal of Wildlife Management 71:2106–2113.

Jennrich, R. I., and F. B. Turner. 1969. Measurement of non-circular home range. Journal of Theoretical Biology 22:227–237.

Johnson, D. H. 1980. The comparison of usage and availability measurements for evaluating resource preference. Ecology 61:65–71.

Keating, K. A., and S. Cherry. 2004. Use and interpretation of logistic regression in habitat selection studies. Journal of Wildlife Management 68:774–789.

Kirk, D. A., A. C. Park, A. C. Smith, B. J. Howes, B. K. Prouse, N. G. Kyssa, E. N. Fairhurst, and K. A. Prior. 2018. Our use, misuse, and abandonment of a concept: Whither habitat? Ecology and Evolution 8:4197–4208.

Klopfer, P. H., and J. U. Ganzhorn. 1985. Habitat selection: Behavioral aspects. Pp. 435–453 in M. L. Cody, ed. Habitat selection in birds. Academic Press, New York.

Koerner, L. 1996. Carl Linnaeus in his time and place. Pp. 145–162 in N. Jardine, J. A. Secord, and E. C. Spary, eds. Cultures of natural history. Cambridge University Press, New York.

Lack, D. 1933. Habitat selection in birds with special reference to the effects of afforestation on the Breckland avifauna. Journal of Animal Ecology 2:239–262.

Lay, D. W. 1956. Effects of prescribed burning on forage and mast production in southern pine forests. Journal of Forestry 54:582–584.

Leclerc, Comte de Buffon, G.-L. 1761. Natural history, general and particular. 1791 translated by W. Smellie. Strahan & Cadell, London.

Leopold, A. 1933. Game management. C. Scribner's Sons, New York.

Leopold, A. 1949. A Sand County almanac. Oxford University Press, Oxford.

Levins, R. 1968. Evolution in changing environments. Princeton University Press, Princeton, NJ.

Linnaeus, C. 1781. Dissertation II, on the increase of the habitable earth. 1781 translation by F. J. Brand, in 1977 reprint of Select dissertations from the Amoenitates Academicae. Arno, New York.

MacArthur, R. H. 1958. Population ecology of some warblers of northeastern coniferous forests. Ecology 39:599–619.

MacArthur, R. H., and R. Levins. 1967. The limiting similarity, convergence, and divergence of coexisting species. American Naturalist 101:377–385.

MacKenzie, D. I., J. D. Nichols, G. B. Lachman, S. Droege, J. A. Royle, and C. A. Langtimm. 2002. Estimating site occupancy rates when detection probabilities are less than one. Ecology 83:2248–2255.

MacKenzie, D. I., J. D. Nichols, J. A. Royle, K. H. Pollock, L. Bailey, and J. E. Hines. 2017. Occupancy estimation and modeling: Inferring patterns and dynamics of species occurrence. Elsevier, Amsterdam, Netherlands.

Mägi, M., R. Mänd, H. Tamm, E. Sisask, P. Kilgas, and V. Tilgar. 2009. Low reproductive success of great tits in the preferred habitat: A role of food availability. Ecoscience 16:145–157.

Manly, B. F. L., L. McDonald, D. L. Thomas, T. L. McDonald, and W. P. Erickson. 2002. Resource selection by animals: Statistical design and analysis for field studies, 2nd edition. Springer Science & Business Media, New York.

Merriam, C. H. 1890. Results of a biological survey of the San Francisco Mountains regions and desert of the Little Colorado River in Arizona. USDA Bureau of Biology, Survey of American Fauna 3:1–132.

Messmer, T. A. 2009. Human-wildlife conflicts: Emerging challenges and opportunities. Human-Wildlife Conflicts 3:10–17.

Mohr, C. O. 1947. Table of equivalent populations in North American small mammals. American Midland Naturalist 37:223–249.

Moore, E. K., G. R. Iason, J. M. Pemberton, J. Bryce, N. Dayton, A. J. Britton, and R. J. Pakeman. 2018. Habitat impact assessment detects spatially driven patterns of grazing impacts in habitat mosaics but overestimates damage. Journal for Nature Conservation 45:20–29.

Morrison, M. L., and L. S. Hall. 2002. Standard terminology: Toward a common language to advance ecological understanding and applications. Pp. 43–52 in J. M. Scott, P. J. Heglund, M. L. Morrison, J. B. Haufler, M. G. Raphael, W. A. Wall, and F. B. Samson, eds. Predicting species occurrences: Issues of scale and accuracy. Island Press, Washington, DC.

Odum, E. P. 1959. Fundamentals of ecology, 2nd edition. Saunders, Philadelphia.

Oliver, F. W. 1913. Some remarks on Blakeney Point, Norfolk. Journal of Ecology 1:4–15.

Pianka, E. R. 1994. Evolutionary ecology, 5th edition. Harper Collins College, New York.

Priestley, J. H. 1913. The quadrat as a method for the field excursion. Journal of Ecology 1:89–94.

Regan, T. 1983. The case for animal rights. University of California Press, Berkeley.

Rosenberg, D. K., and K. S. McKelvey. 1999. Estimation of habitat selection for central-place foraging animals. Journal of Wildlife Management 63:1028–1038.

Royle, J. A., R. B. Chandler, R. Sollmann, and B. Gardner. 2013a. Spatial capture-recapture. Academic Press, New York.

Royle, J. A., R. B. Chandler, C. C. Sun, and A. K. Fuller. 2013b. Integrating resource selection information with spatial capture-recapture. Methods in Ecology and Evolution 4:520–530.

Royle, J. A. and J. D. Nichols. 2003. Estimating abundance from repeated presence-absence data or point counts. Ecology 84:777–790.

Royle, J. A., J. D. Nichols, and M. Kéry. 2005. Modelling occurrence and abundance of species when detection is imperfect. Oikos 110:353–359.

Schaus, W. 1899. New species of Lithosiidæ from tropical America. Journal of the New York Entomological Society 7(3):214–217.

Schmidtz, D., and E. Willott. 2002. Environmental ethics. Oxford University Press. New York.

Schoener, T. W. 1974. Resource partitioning in ecological communities. Science 185:27–39.

Schoener, T. W. 1989. The ecological niche. Pp. 79–113 in J. M. Cherrett, ed. Ecological concepts. Blackwell Scientific, Oxford.

Singer, P. 1990. Animal liberation, 2nd edition. Avon, New York.

Stauffer, D. E. 2002. Linking populations and habitats: Where have we been? Where are we going? Pp. 53–61 in J. M. Scott, P. J. Heglund, M. L. Morrison, J. B. Haufler, M. G. Raphael, W. A. Wall, and F. B. Samson, eds. Predicting species occurrences: Issues of scale and accuracy. Island Press, Washington, DC.

Stoddard, H. L. 1931. The bobwhite quail: Its habits, preservation, and increase. Charles Scribner's Sons, New York.

Stone, C. D. 1974. Should trees have standing? Toward legal rights for natural objects. William Kaufmann, Los Altos, CA.

Stone, C. D. 1987. Earth and other ethics. Harper & Row, New York.

Svärdson, G. 1949. Competition and habitat selection in birds. Oikos 1:157–174.

Swihart, R. K., and N. A. Slade. 1985. Influence of sampling interval on estimates of home-range size. Journal of Wildlife Management 49:1019–1025.

Thomas, J. W., and D. E. Toweill, eds. 1982. Elk of North America. Stackpole, Harrisburg, PA.

Thorpe, W. H. 1945. The evolutionary significance of habitat selection. Journal of Animal Ecology 14(2):67–70.

USFWS [US Fish and Wildlife Service]. 2003. 2001 national and state economic impacts of watching wildlife. US Fish and Wildlife Service, Washington, DC.

Van Balen, J. H. 1973. A comparative study of the breeding ecology of the great tit Parus major in different habitats. Ardea 61:1–93.

Van Horne, B. 1983. Density as a misleading indicator of habitat quality. Journal of Wildlife Management 7:893–901.

Verner, J., M. L. Morrison, and C. J. Ralph, eds. 1986. Wildlife 2000: Modeling habitat relationships of terrestrial vertebrates. University of Wisconsin Press, Madison.

Wade, A. A., K. S. McKelvey, and M. K. Schwartz. 2015. Resistance-surface–based wildlife conservation connectivity modeling: Summary of efforts in the United States and guide for practitioners. General Technical Report RMRS-GTR-333. US Department of Agriculture, Forest Service, Rocky Mountain Research Station, Fort Collins, CO.

Wallmo, O. C., ed. 1981. Mule and black-tailed deer of North America. University of Nebraska Press, Lincoln.

Werner, E. E., J. F. Gilliam, D. J. Hall, and G. G. Mittelbach. 1983. An experimental test of the effects of predation risk on habitat use in fish. Ecology 64:1540–1548.

Werner, E. E., and D. J. Hall. 1979. Foraging efficiency and habitat switching in competing sunfishes. Ecology 60:256–264.

Wiens, J. A. 2016. Ecological challenges and conservation conundrums: Essays and reflections for a changing world. John Wiley & Sons, Hoboken, NJ.

Woodhouse, S. W., G. A. McCall, E. Hallowell, J. L. LeConte, S. F. Baird, and C. Girard. 1852. Descriptions of new species of birds of the genera Vireo, Vieill., and Zonotrichia, Swains; Note on Carpodacus frontalis, with description of a new species of the same genus, from Santa Fé, New Mexico; Description of new species of Reptilia from Western Africa; Remarks on some coleopterous insects collected by S. W. Woodhouse, M.D., in Missouri Territory and New Mexico; Characteristics of some new reptiles in the Museum of the Smithsonian Institution. Proceedings of the Academy of Natural Sciences of Philadelphia 6:60–70.

Worton, B. J. 1989. Kernel methods for estimating the utilization distribution in home range studies. Ecology 70:164–168.

Yapp, R. H. 1922. The concept of habitat. Journal of Ecology, 10:1–17.

3 — Heterogeneity and Disturbance

But the subjects of these actions—landscapes and the biota or resources they contain—are not members of the same cultural system as humans, and our goals or values may often be irrelevant or antithetical to their well-being or persistence. They may not play the game according to our rules.

Wiens (2016:89)

The habitat of a species is defined by a location and the set of conditions used and selected by members of this population to meet some or all of their life needs (Chapter 1). Components of habitat have been classically viewed as consisting of food, water, and cover (Leopold 1933; Edminster 1938), and, more recently, as including a wide variety of other environmental factors. Habitat encompasses a very broad array of resources that become useful to an individual at different times. Hence the temporal component of habitat descriptions becomes critical in assessing population viability.

Consider the well-studied habitat needs of Rocky Mountain elk (*Cervus canadensis nelsoni*). Common management understandings are that this species needs a mixture of forage and cover, which requires an examination of broad spatial extents (i.e., landscape; see below). But the needs of elk are much more complex. Consider a dense stand of mature trees, which managers would call thermal cover. These trees will only serve as such if they are suitably adjacent to forage areas and water sources. Forage needs for elk can be met in an area with both wet meadows, which are likely to remain green throughout the summer, and sheltered semi-arid bunchgrasses, which would receive little snow and therefore provide forage during the winter. Further, the

herds must be protected from overhunting, which might involve controling human access, such as by road closures and hunting regulations that limit the number of elk to be harvested each season. If all of these features are in place, then the elk may be able to use the stand as thermal cover and, thus, as part of their habitat.

It is the dispersion of all such habitat elements across space and over time that, in part, determines the distribution, density, and demographics of wildlife populations. In most situations, disturbance dynamics are a primary force in shaping the size and juxtaposition of features in the environment. Thus habitat elements and the disturbance processes that create or modify them are inexorably mixed. In this chapter, we focus on concepts and measures of habitat at broad spatial extents through space and over time and look at the processes that generate them.

In Chapter 1, we have introduced the idea that landscapes are embedded within the broader ecosystem concept. Landscapes represent a physical attribute of the ecosystem and are useful in describing how structural features of interest are arranged through space and over time within a particular area for the relevant species' population. Because we focus on the population level, we can view a landscape perspective as what is being perceived collectively by members of a biological population, along with the critical time component. Throughout the literature (and as has been developed in Chapter 1), we see that landscape has been viewed largely from a human perspective—namely, as a large area delineated by researchers, rather than one explicitly relative to the occurrence, use, and selection by individual animals and populations. That said, there is much to be gleaned from the literature on how animal behavior and, ultimately, persistence is tied back, over time, to broad-scale heterogeneity in the environment. Below, we use landscape in the context of that species-specific, wide spatial extent that a population will encounter, either in part or throughout its annual cycle.

Why Study Habitats and Animals in Landscapes?

Landscapes are the great environmental integrators. Within them, individuals interact as populations, and different species interact and overlap with abiotic elements of ecosystems. Like habitat, landscapes are defined by the organisms that use them. For a Laysan albatross (*Phoebastria immutabilis*), its landscape covers thousands of square kilometers of ocean; for wolverines (*Gulo gulo*), perhaps a mountain range. For invertebrate micro- and mesofauna that contribute critically to soil productivity, their landscape is composed of interstices between soil particles. Within the landscape of a species, dynamics and disturbance regimes interplay with its ever-changing biota, across scales ranging from sweeping catastrophic fires and volcanic eruptions to soil pits and sun flecks created from single treefalls. As wildlife researchers or managers, we often are challenged to predict the responses of environments, organisms, and ecosystems to our activities within landscapes.

Working with landscapes poses special problems for researchers and managers. The former must contend with studying subjects that usually have no or few replicates and controls, limited baseline (predisturbance) data, and all the environmental noise that complicates data analysis and confounds interpretation. The latter must cope with managing in the face of great variability and uncertainty—including changeable ecological conditions and disturbance events across wide scales of space and time.

A landscape approach to assessment and management of wildlife-habitat relationships is nevertheless useful, because some ecological processes emerge at broad spatial scales that are not evident or cannot be easily understood within smaller areas, including the biogeography of species distributions; hydrology of surface and subsurface flows; pedology of soil formation, change, and erosion; as well as population dynamics and species exchanges, invasions, speciation, and extinction. Humans also occupy the landscapes of different populations and use resources that can have drastic effects on, and be affected by, each of these processes.

Definition and Classification of Landscapes

A landscape is an area of the earth's surface that consists of a complex of systems, formed by the activity of abiotic and biotic factors (including people), which, by its physiognomy, forms a recognizable entity (Zonneveld 1979). In this general definition, landscapes do not necessarily have a specific size and might be considered at various scales, depending on the area over which unique constellations of ecological processes operate to form recognizable entities. Moreover, here landscape is borderless and, while not arbitrary in nature, carries the same weaknesses as other terms that have been discussed in Chapter 1 (e.g., community). We do need to consider the conditions that an organism encounters as it moves through space and over time and assess how those conditions are likely to change. Our point is that the word "landscape" and its related concepts must be carefully defined and, to the extent possible, articulated in terms of their spatial extent, even if the boundaries are somewhat vague. We tend to view landscapes in our own terms of spatiotemporal scale, however (see Chapter 1). A study that occurs at the landscape scale generally will not connote a study at the spatial scale of soil particles, or at the temporal scale of the Holocene. But these are simply anthropocentric understandings and do not work well to provide information on the fundamental concept of landscapes.

In a more effective definition, landscapes can be described as containing a common set of elements, including plant and animal communities, ecological processes, and disturbance regimes, all of which coincide in a particular geographic setting, although individual elements may extend beyond the delineated landscape area. A landscape will, over the long run, be stable as long as these processes and disturbances remain constant.

Changes in landscapes, however, are often synergistic. A good example is found in the spread of exotic annual grasses worldwide. These grasses are superb competitors in arid environments and lead to profound changes in the areas they invade (Schnupp and Delaney 2012). Additionally, they thrive on fire and, being annuals, senesce early in the season, providing a contiguous layer of fine fuels that is perfect for the promulgation of large, fast-moving fires. For this reason, their presence also alters the fire regime (D'Antonio and Vitousek 1992), often from one where there were few large fires, due to fuel limitations, to one in which large fires are frequent (Masters and Galley 2007). This changes the vegetation composition and structure (e.g., patchiness) and profoundly affects ecological processes. (We discuss dynamics of disturbance further below.)

The landscape concept may be applied to a broad variety of environments, including marine systems. Although we focus on terrestrial systems in this chapter, the concepts can be applied to aquatic situations. A coastal, estuarine, lacustrine (lake), or even oceanic environment may contain recognizable entities or landscapes if we simply look at the right elements. For example, cliff-nesting seabirds may be described as inhabiting coastal landscapes, which consist of recognizable entities defined by climatic

and physiographic components. Even pelagic environments used by albatrosses and whales can be described as marine landscapes (or waterscapes)—defined by current cells, air circulation patterns, upwelling patterns, benthic topography, and weather systems—despite the fact that we cannot readily see these elements without measuring devices or help from satellite imagery. Understanding the dispersal of organisms in aquatic, particularly marine, environments may aid explanations, and thus the management, of these areas (Kinlan and Gaines 2003).

Landscape Ecology

Understanding the patterns, processes, and reasons for the distribution of species and landscape elements is the challenge of landscape ecology. Under our operating paradigm (Chapter 1), the pertinent biological population delineates the physical boundary within which we focus our studies of habitat. The concept of landscape defines that boundary and adds a time element. Individuals within a population usually do not occupy every part of the landscape throughout time. Hence the landscape concept not only incorporates aspects that define habitat, but also addresses how the relevant structure and processes change over time throughout the area bounded by that population. Thus a functional definition of landscape ecology is the study of the responses of organisms, species, ecological communities, and ecosystem processes to each other in spatially heterogeneous and temporally changeable environments across a broad area. When viewed in the context of a single species, landscapes are delineated by that species' population.

Forman and Godron (1986), while not considering landscapes in the formal species-centric approach we apply here, proposed a useful system for landscape classification. They broadly categorized terrestrial landscapes into 6 general types, based on the degree of human disturbance: natural, managed, cultivated, suburban, urban, and megalopolis. Natural landscapes feature perhaps special conditions that provide sources of refugia for species sensitive to human disturbance and other stressors, as well as afford benchmarks from which to measure changes in biodiversity, population dynamics, and ecosystem processes affected by human intervention. Managed and cultivated landscapes, including managed forests, livestock grazinglands, and agricultural lands, collectively constitute what are more frequently being called "semi-natural landscapes" (e.g., Soderstrom and Part 2000). Suburban and urban landscapes, although not considered to be natural or semi-natural, still provide parks, greenbelts, forest remnants, tree cover, and other elements studied in the discipline of urban wildlife management (Mortberg 2001). Indeed, animals are increasingly adapting to urban environments (Luniak 2004), and these locales achieve high levels of both density and diversity (Crooks et al. 2004). The megalopolis landscape offers perhaps the greatest challenge to habitat managers, as it consists of geographic areas in which much of the native biota is lost and where mostly exotic species thrive.

Scales of Ecological and Management Issues

One of the major strengths in the discipline of landscape ecology is integrating ecological studies across scales of space and time. The term scale, however, is often used ambiguously in landscape ecology (and most other) studies, and it begs a clarity of definition (Peterson and Parker 1998; Withers and Meentemeyer 1999; Jenerette and Wu 2000). In Table 3.1, we list 6 dimensions of scale, along with suggestedguidelines for 3 levels of magnitude for each dimension.

Geographic extent is 1 dimension of scale. Typically, the term "large scale" is used in landscape studies to connote a large geographic extent—which generally means big, from the human viewpoint. This term is unfortunate, because in cartography, the same term refers to larger values of map-scale ratios—that is, the ratio of a fixed distance on a map, such as 1 inch, to the distance of the actual geographic surface it represents, Thus, in cartography, large-scale equates to a small geographic extent. Mapping ecological entities, such as soil map units (see Valentine 1981) can vary by map scale. Thus it is important to denote this scale. Moreover, unless the context is specifically cartographic, we also suggest replacing the ambiguous terms "large-scale" with "large geographic extent" and "small-scale" with "small geographic extent," or, better yet, simply stating the geographic extent. We think that many, if not most, ecologists remain vague in their use of spatial scale, because they do not address the boundary problem. Rather, they fall back on largely borderless terms (e.g., "a study population").

Another dimension of scale is spatial resolution. Unlike geographic extent, which is a physical property of landscapes, spatial resolution refers to the subjective level of detail with which we choose to evaluate the landscape. For the most part, spatial resolution is not driven by the biological grain (Weins 1976) of the organisms of interest or by the natural grain of disturbance processes, but rather by technological limitations, such as the constraints of satellite sensor arrays and the abilities of our computers to assimilate data (Weins 1989). Spatial resolution can refer to a variety of images, ranging from coarse-grained pixels or large vector polygons to those with very fine-grained resource patches or point locations of conditions. Whereas geographic extent and map scale are roughly (inversely) correlated—that is, large-scale maps tend to cover smaller geographic extents (depending on the physical size of the map)—spatial resolution can be a more independent dimension. That is, a small-scale map (e.g., 1:2,000,000 scale) covering a major drainage basin (e.g., the St. Lawrence Seaway watershed), might be represented by coarse-grained spatial resolution (e.g., 1 km^2 pixels from remote sensing satellite images), or by finer-grained spatial resolution (e.g., vector polygons <10 ha). Thus it is important to denote the spatial resolution, as well as the geographic extent and cartographic map scale, of a particular map image (Salajanu and Olson 2001).

Table 3.1.   Various aspects of scale and examples at 3 levels of magnitude

Aspect of scale	Broad-scale magnitude	Midscale magnitude	Fine-scale magnitude
Geographic extent	Entire major drainage basins, or entire ecoregions (large-scale study)	Subbasins or local watersheds, or more local physiographic provinces; large groups of vegetation patches (medium-scale study)	Areas smaller than subbasins or local watersheds; small groups of individual vegetation patches or substrates ("small-scale" study)
Map scale (cartographic)	Small-scale maps (e.g., ≥1:1,000,000)	Medium-scale maps (e.g., 1:100,000)	Large-scale maps (e.g., 1:24,000)
Spatial resolution	Typically coarse-grained, such as for characterization of the vegetation and environmental conditions at 100 ha pixel size	Environmental patches within subbasins or local watersheds (e.g., ≥10 ha)	Fine-grained (e.g., <10 ha patch sizes)
Time period	Paleoecological past and evolutionary future	Historical past and approximately 1 century into the future	Very recent management past and current conditions projected only a few years into the future
Administrative hierarchy	International treaties, such as on biodiversity and plant and animal commerce; also national laws and land management or resource regulations (the strategic scale)	Individual agency or tribal-specific policies and legal guidelines; state- or provincial-level agency policies, industry policies, and private landholder resource management goals	Local management's unit-level operational policies, down to project-level operations (the tactical scale)
Biological organization	General abundance of vegetation communities, cover types, and structural stages; mapped locations of ecoregions or ecosystems; inference to broad ecological communities and species assemblages (coarse-filter elements)	Distribution and abundance of individual species or species groups	Species; gene pools; subspecies or varieties; morphs or ecotypes (fine-filter elements)

Time period is another dimension of scale that should be made explicit in ecological studies. Specifying time periods and study durations is vital to correctly interpreting geographic data and landscape simulations (Meentemeyer 1989), particularly for disturbance events and responses by organisms and populations. Ideally, disturbances should be empirically measured by their duration and frequency, in order to best interpret recurrent patterns and potential influences from human activities. Also, it is important to report the time period over which data were gathered for a geographic analysis (or a particular map) in landscape studies, as some ecosystem processes, including disturbance regimes, may occur

beyond the time period studied. We discuss disturbances in more detail below.

Another dimension of scale, administrative hierarchy, is more pertinent to management use but is not often explicitly addressed. It refers to the breadth of political, social, cultural, or even economic mandates and policies of governments, which may play important roles in some studies or management plans, as well as in interpreting observations (Herzog et al. 2001). Particularly for management, it could be useful to explicate which organizational hierarchy levels apply. For example, at a broad-scale magnitude, studies of marine landscapes might pertain to international fishing treaties beyond the coastal sovereign fishing zones of individual countries, and they may also involve national policies for the protection of marine mammals, fish stocks, or coral reefs. It could be important to clarify which scales of legal mandates and resource management policies pertain.

Lastly, a level of biological organization refers to the biological dimension of scale—that is, to whether a study or plan pertains to ecosystems, communities, assemblages, species, or more finely defined entities, such as populations, gene pools, or ecotypes (see Chapter 1). This dimension could also refer to classification levels of vegetation communities or ecosystems, such as plant associations, types of vegetation, and ecoregions (e.g., Bailey 2005).

Since landscapes should be biologically defined entities, the level of biological organization delineates the scale at which precision is needed and from which all other scales naturally follow. The scale of biological organization will imply a necessary geographic extent and an optimal spatial resolution and temporal domain. Failure to properly define this scale will lead to a cascade of incorrect decisions throughout the study. As such, describing a management plan as ecosystem management (e.g., Armstrong et al. 2003; Butler and Koontz 2005)—a management paradigm that has encompassed most federally controlled wildlands in the United States—is confusing and largely meaningless without a clear understandings of precisely what is being managed, and at what spatiotemporal scales. Although we focus on a biological population as the fundamentally most relevant level for investigation, all studies—and thus management plans—should include an explicit discussion of this topic.

When thinking about the appropriate scales with which to study a landscape, it is important to understand that domains of space and time do not automatically connote coarse resolutions or broad levels of biological organization. A landscape study or management plan might pertain to a fine-scale magnitude of biological organization, such as an inventory of ecotypes, that occurs across a broad geographic

extent, such a drainage basin. In this way, the various dimensions of scale may be applied at different magnitudes for a given purpose. Overall, it is important to describe these various dimensions of scale for products based on maps and remote-sensing images, in order to explain the context in which the information should be used. Applying a geographic information system (GIS) model at the wrong scales of geographic extent, resolution, and so forth can lead to false conclusions. Likewise, understanding the dimensions of scale of a geographically based analysis, such as mapping hot spots of species richness, helps guide the scales at which such information should and should not be used. For example, species-richness maps depend on the scale (geographic extent and spatial resolution) of the mapping effort (Stoms 1994).

In conclusion, we suggest that habitat and landscape studies (and management plans) clearly identify the magnitudes of geographic extent, map scale, spatial resolution, time period, organizational hierarchy (if appropriate), and level of biological organization to be addressed and evaluated. In this way, considerable confusion over terms and methods can be avoided. Scientists and managers need to be much more explicit about describing the boundaries of their work and its applicability in space and time.

Scales of Management Authority and Considerations

As noted above, landscapes come in a wide range of spatiotemporal scales, depending on the organisms and processes of interest. Being people, however, we usually think in terms of human scales—spatial scales that make sense to us and time scales measured in human lifetimes—to describe landscape-scale studies (both specifically and more generally) in the field of ecology. Investigations with larger or smaller spatiotemporal scales, while still examining landscapes, have generally been viewed as separate disciplines. Figure 3.1 represents a space-time zoning map that denotes various scientific disciplines

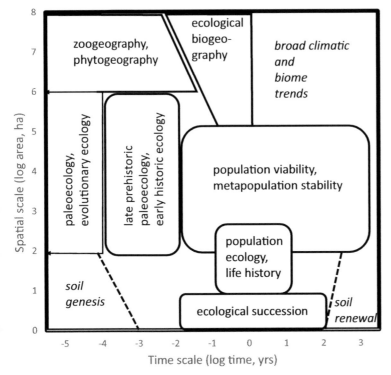

Figure 3.1 A "zoning map" of ecological scale and species assessment, plotting spatial area against time (note the log axes). This figure illustrates the spatial and temporal extents of various disciplines that may be useful in wildlife-habitat management and land use planning. (Reproduced from Morrison et al. (2006), with the permission of Island Press)

useful in landscape ecology studies. For example, the upper left corner represents studies of the biogeography of animals and plants (zoogeography and phytogeography, as well as phylogeography, sensu Avise 2000), roughly over a time period covering the recent past to the distant geologic past, and over a spatial extent of approximately 10^6 to 10^8 ha. Studies of ecological biogeography might reach from the recent past to the near future, and perhaps involve smaller spatial extents, say down to 10^5 ha. Researchers of climate change are interested in projecting global or regional climatic and biome trends from the present to perhaps centuries or a millennium into the future, over approximately the same spatial extent.

At a smaller temporal scale, studies of paleoecology and evolutionary ecology typically pertain to 10,000 years in the past and further back, whereas late prehistoric paleoecology (e.g., of Holocene events) and early historic ecology may peer back only a century to 10 millennia or so ago, over a spatial ex-

tent of perhaps a hundred to a million hectares, depending on the specific study. An analysis of population viability, including the stability of metapopulations (see Chapter 4), might apply to hindcasting a few decades to a century into the past and forecasting decades to a few centuries into the future. Its spatial extent is often less than that of paleoecological studies, again depending on the range of the species and environments under consideration.

Studies of population ecology and life histories of species typically are concerned with the recent past and the very near future, perhaps on the order of just a few decades or a few generations of the species of interest, and cover more-localized geographic extents. Studies of ecological succession usually are site specific, but they can address conditions a century or so in the past and a century or so into the future. Finally, pedogenesis (soil origin) typically occurs or is studied on a mostly site-specific basis, about a millennium or more into the past, whereas prospects for soil renewal may not begin for centuries into the future. Other disciplines also can be added to the zoning map.

In addition to organizing scientific thought, this type of space-time zoning map may be useful for identifying realms of duration and geographic extent over which managers should address planning effects. This is shown in Figure 3.2, where ecosystem management issues are plotted, corresponding to the disciplines listed in Figure 3.1. In the upper left corner, at time periods in the distant past and areas of large geographic extent, a manager might need to understand the geographic origins of species, including their centers of origin, the centers and routes of their spread, and the role of refugia. Such understandings, going beyond simply academic interest, can greatly help identify locations of evolutionary significance that deserve conservation consideration. An example can be found in the complex Alexander Archipelago of southeast Alaska and northern British Columbia, where unique subspecies and species have evolved and paleoecological conditions have brought large carnivores into sympatry. If ecosystem managers want to retain the range of current broad-scale conditions, then they should look to the recent past and near future across wide geographic areas. And, still across a broad geographic extent, a manager might peer into the near future (a century to a millennium) to project climatic or biome effects when considering such potential influences as climate change, acid precipitation, and ozone depletion.

Over a lesser geographic extent, a manager might need to describe evolutionary, late prehistoric, early prehistoric and early historic, or recent historic conditions. This would be useful for determining the range of natural conditions in an area, as well as for discovering if historical conditions truly represent environments in which species persisted over the long term, or even evolved. One example is a study of the macroecology and paleoecology of the interior Columbia River Basin in the inland west of the United States, which provided managers with a perspective of how recently the current distribution of fauna was influenced by Late Quaternary conditions of climate, vegetation, and associated species' extinctions and originations (Marcot et al. 1998).

The goal of ensuring viable populations should prompt a manager to look into the future (on the order of a century to a millennium), if projections allow, and over a geographic extent of perhaps up to half a million hectares or so, depending on the environment and species of interest. Concern for near-term harvest levels for game animals should be nested within broader desires for maintaining long-term sustainability and viability of the target species, but these aspects pertain to only the near future and cover small geographic areas (Sands et al. 2013).

Management for long-term productivity, diversity, and sustainability of ecosystems can draw from all of these scales and issues. A description of soil genesis is a site-specific attribute that looks back a millennium or more in the past, and an assessment describing long-term soil protection should look centuries to millennia into the future.

Time-space perspectives can aid land management planning and influence the degree to which such efforts might study the past and project future cumulative effects. Most land-resource management plans are designed to operate for perhaps a decade. Future revisions would come later, as part of a subsequent cycle. Yet even if the formal duration of a plan is short, one should peer back in time and project into the future to better understand historical conditions and predict future long-term effects, lest we continue to chip away at our resource base by what may be called the long-term tyranny of short-term actions.

Some management issues ought to be addressed at each planning level, and specific scientific disciplines should be brought to bear in informing managers about them. For example, in the United States, a plan developed for an entire national forest or national grassland may formally be in effect for a decade or so and pertain to an area on the order of 10^4 to 10^5 ha. Issues that the plan would proximately deal

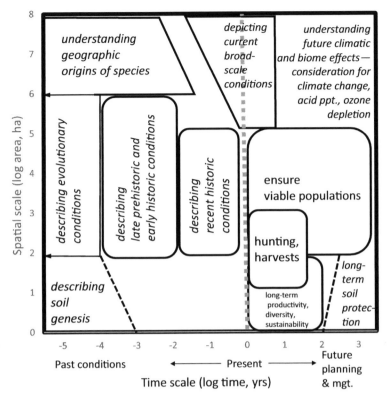

Figure 3.2 A "zoning map" of ecological scale and ecosystem management issues, plotting spatial area against time (note the log axes). Compare it with Figure 3.1, which suggests some of the issues that managers may wish to address in wildlife-habitat plans. (Reproduced from Morrison et al. (2006), with the permission of Island Press)

with (Fig. 3.2) might include ensuring viable populations and, to some extent, representing the full range of ecological or environmental conditions there. In constructing the plan, however, managers should peer into the past and project cumulative effects into the future. With regard to the past, they need to describe late prehistoric, early historic, and recent historic conditions, through the use of paleoecology and early historic ecology. To ensure future viable populations, they might employ a population viability analysis.

It is our view that an ecosystem management approach should address the full array of policies and land and resource management plans, tiered across all geographic extents that account for past and current conditions, as well as any future cumulative effects of human activities on biotic and abiotic conditions. Further, such a sweeping plan would embody

consistent implementation standards, management goals, evaluation criteria, inventory and monitoring activities across all geographic extents, levels of biological organization, and administrative hierarchies ranging from national policies to local projects. In this way, all current issues and conditions, as well as past conditions and future effects, can be consistently treated and monitored across the full breadth of management issues and the cumulative effects of past, current, and future activities.

This discourse on planning at multiple scales illustrates that human influences, as well as natural disturbance regimes, can alter the types, spatial patterns, and development of wildlife habitats. In the next section, we explore specific ways to gauge and interpret heterogeneity of habitats; after that, we discuss disturbance dynamics.

Habitat Heterogeneity
Definitions and Concepts

We define "habitat heterogeneity" as the degree of discontinuity in environmental conditions across a landscape for a particular species. The heterogeneity of resource patches in landscapes has been discussed by different authors in various ways (Table 3.2). Among the aspects of resource patch heterogeneity are patch type richness and diversity; patch dynamics, connectivity, and isolation; fragmentations of environments; vegetation corridors; and edge permeability, effects, and contrast. Modern landscape ecology is based on the patch mosaic paradigm, in which landscapes are conceptualized and analysed as mosaics of discrete patches, and we follow this understanding here. It should be noted, however, that there exist continuous models for describing landscape heterogeneity (McIntyre and Barrett 1992; Manning et al. 2004; McGarigal and Cushman 2005; McGarigal et al. 2009). Since these approaches have not achieved widespread use, we confine our discussion here to patch-based metrics.

When evaluating habitat heterogeneity, it is important to remember that habitat is a species- or organism-centric term, and that a particular environmental condition or gradient may constitute habitat for 1 species and a barrier for another. Thus, though often related, habitat heterogeneity does not equate to patch heterogeneity, a term generally associated with vegetation or edaphic patterns.

Environmental conditions can include vegetation composition and structure, as well as dynamic flows of energy, nutrients, resources, and fluids (water and air). It additionally can include population patterns, such as prey or predator densities, as well as anthropogenic factors, such as the presence of domestic grazing. Discontinuities in environmental conditions occur as ecotones (relatively sharp breaks in environmental conditions) or ecoclines (broader gradations in conditions over areas of greater geographic extent). Discontinuities can occur naturally, such as with changes in soil types or edges of water bodies, or anthropogenically, such as with edges of plowed grasslands or burned forests.

Fragmentation

In this book, fragmentation refers to the degree of heterogeneity of a habitat that serves to separate or isolate resource patches. Fragmentation is most commonly mapped with patches of vegetation or other necessary resources, such as water bodies. The size distribution, degree of separation, and shape of resource patches are all components of habitat fragmentation. Since it refers to habitat, fragmentation is necessarily a species-specific condition. Unfortunately, the term "habitat fragmentation" is engrained in the ecological literature and used to refer to virtually any sort of heterogeneous condition (most often a more appropriate term would be "vegetation fragmentation"). Authors thereby equate fragmentation with species extirpation (Gu et al. 2002; Solé et al. 2004) in a far too general way. Also, many of them (e.g., Bogaert et al. 2005) refer to "landscape fragmentation," which is, strictly speaking, incorrect, since it is environments or resources (habitats for specific species) that become fragmented within landscapes, not entire landscapes per se.

Habitat-fragmentation analyses associated with the evaluation of identified patches on a landscape are problematic, due to 2 major issues. The first is that fragmentation is a categorical property. That is to say, statistically it requires a finite group of things that can be evaluated in terms of their sizes, spatial juxtapositions, and shapes. As such, it requires that data (in virtually all cases vegetation data) be grouped into classes, which then resolve into patches of specific classes juxtaposed on a landscape. The second is that vegetation fragmentation does not equate to habitat fragmentation.

Catagorization of Fragmentation
To understand the first issue, one needs to recognize that the categorization of a landscape into discrete classes of patches is largely arbitrary. Consider a par-

Table 3.2.   Components of habitat heterogeneity in landscapes. Papers listed as sources are seminal works where concepts associated with specific components were elucidated. In addition, Neel et al. (2004), Cushman et al. (2008), and Šímová and Gdulová (2012) provide more-recent reviews. For a detailed discussion of the plethora of associated metrics, see McGarigal (2015).

Component	Description	Example	Source
Patch type richness and diversity	Number and relative area of habitat types for a species within a landscape	Number of patches of different types affects species richness, diversity, and numeric dominance of a species in a community	Bascompte et al. 2002, Fitzsimmons 2003
Patch dynamics	Incursion and melding of patches over time as a function of disturbance events and the successional growth of vegetation	Distribution of vegetation patches over time as affected by stand-replacing fires and subsequent regrowth	Wu and Loucks 1995
Patch connectivity	Degree of adjacency of patches with similar conditions in a landscape	Connectivity of fencerow edge habitat in an agricultural setting, differentially affecting wildlife with different long-range dispersal capabilities	Schooley and Wiens 2003, Walker et al. 2003
Patch isolation	Distance from 1 type of patch to the next (or nth) nearest patch of the same type	Isolated patches less often colonized by species that do not disperse easily through unsuitable environments, such as Bachman's sparrow (*Aimophila aestivalis*) in South Carolina pine woodlands	Bender et al. 2003, Tischendorf et al. 2003
Fragmentation	Breakup of contiguous environmental or habitat patches into smaller, more disjunct, and more isolated patches of different types	Fragmentation of grasslands differentially affecting bird species with varying sensitivity to habitat areas; differential effects of forest fragmentation on the population viability of arboreal marsupials in Australia	Silva et al. 2003, Tallmon et al. 2003
Vegetation corridors	More-or-less linear or constricted arrays of environments or habitats in a landscape, serving to connect larger patches	Movement corridors in undisturbed riparian woodland for cougars (*Felis concolor*) in southern California mountains; riparian woodland and shelterbelt corridors in North Dakota, supporting populations of migratory birds	Mabry and Barrett 2002, Haddad et al. 2003
Edge permeability	Degree to which an organism can move among patches within a landscape	Effects of microhabitats and microenvironments within clearcuts on the dispersal of red-legged frogs (*Rana aurora*)	Stamps et al. 1987, Chan-McLeod 2003
Edge effect	Incursions of microclimate and vegetation into a patch, typically forested, from a disturbed edge or opening	Clearcuts causing reduced tree-stocking density, increased growth and reproduction of dominant trees, higher tree mortality, and incursion of warmer, drier microclimates into adjacent old-growth forests	Chen et al. 1992, 1995
Edge contrast	Degree of difference in the vegetation structure between 2 adjacent patches	Great contrast in average daily air and soil temperatures, wind velocity, short-wave radiation, and air and soil moisture, all of which differ significantly between clearcuts and old-growth forests	Chen et al. 1993

tially forested landscape. One could categorize it into forest and non-forest, or one could classify it as deciduous forest, coniferous forest, brush, grassland, and other—and these are just 2 of a vast number of potential categorization schemes. Further, there is the question of how one determines whether an area of the landscape is in 1 class or another. What is forest, exactly? How many trees per hectare represent the minimum density? When does a regenerating stand cease to be brush and become forest? There are no correct answers to these questions. Largely for this reason, aerial photo interpretation was abandoned as the main information source for analyzing habitat fragmentation, due to failures to delineate landscape features in a repeatable manner (e.g., Sadowski et al. 1987).

With most modern, remotely sensed data, landscapes are represented by pixels, which are uniform areas of some arbitrary size that generally have the property of light reflectance. The spectrum of the reflected light associated with each pixel is used to categorize it, sometimes adding additional information, such as elevation and aspect, to the algorithm (e.g., Bahadur 2009). Pixels are grouped into classes, based on their similarity to other pixels. Broadly speaking, there are 2 approaches to classifying pixels: unsupervised, in which pixels are simply clustered, based on their similarity; and supervised, where pixels are compared with an established set of pixels of known types and grouped with the type with which they are most similar (Khorram et al. 2016). For example, one would identify areas of forest through an on-the-ground examination and choose pixels from these known forest locations to serve as archetypes for forest classification. The classification of all other pixels is then based on their spectral similarity to the set of archetypes (called a "training set"). Supervised classification critically depends on the idea that spectral similarity to a pixel associated with a land-cover type is likely to represent that same type (Foody 2002), a property that is generally assessed by basing it on a confusion (or error) matrix that depicts the frequencies of correct and incorrect categorization.

Although the classification approach differs from methods based on identifying objects (e.g., trees), it should be clear that both supervised and unsupervised classification-scheme approaches have done nothing to address the problems associated with deciding what is to be considered as a type and how many types should be categorized. Moreover, they have additional artifices, due to the pixel representation itself. As mentioned above, a pixel is a single homogeneous object that has reflectance properties. Because pixel-level homogeneity is assumed, it really matters how large a pixel is. Landsat and MODIS are 2 satellite platforms with imagery that is commonly used for landscape evaluation, but they have very different spatial resolutions. In Landsat imagery, pixel sizes are roughly 30 m on a side (or $\sim$900 m^2). MODIS pixels are approximately 500 m on a side (or $\sim$250,000 m^2). Over 250 Landsat pixels could fit into a single MODIS pixel. The likelihood that a MODIS pixel–sized area, if classified based on Landsat imagery, would be considered to be a single type is low for most landscapes. Therefore the classification of the landscape will differ, due to pixel size alone.

For an example of how this difference affects landscape analysis, consider an area covered by a single MODIS pixel that is 30 percent forest and 60 percent grassland. Evaluated as a single pixel, this area would be classified as a non-forest type, because the majority of its spectral signature is associated with grasses. If, however, that same area were categorized using Landsat imagery, a good deal of the forest would be classified as such and separated from the grasslands.

In tests, simple metrics (e.g., the area of a type, such as forest) nearly always change when you alter the pixel size, sometimes spectacularly. If you use a majority rule to classify pixels, rare types tend to vanish when pixel size increases—they are never common enough to classify the entire pixel as that particular type. If you were to classify pixels based on any portion of the area belonging to a type (e.g., a burned pixel being any pixel with elevated infrared emissions), rare types can increase with escalating pixel size (e.g., Hlavka and Livingston 1997).

These problems also interact with levels of vegetation fragmentation. If bits of a minority type are clumped together, those clumps may still be correctly classified, even when pixels are large, but if a rare type is diffuse, it will simply vanish from the landscape map. The proportion of correct classifications also is affected by fragmentation, because landscapes with more edges have more pixels that fall onto these edges, leading to mixed spectral signatures that do not conform to any part of the training set. Some of these issues can be addressed through fuzzy classification, which allows subpixel features to be retained. This can greatly improve the ability to associate animal location data with rare types of vegetation (e.g., McKelvey and Noon 2001). Most fragmentation indices, however, require hard classification.

Because of these issues, landscape fragmentation, at least as we measure it, does not exist outside of the context in which it was created. We can debate whether a landscape can ultimately be referred to as generically fragmented, but we can state without caveats that the degree of fragmentation, as measured by any fragmentation index, is specific to the spatial resolution, map type (e.g., raster or vector), nature of the classifying algorithm and incorporated data, and specific classification rules. Change any of these, and the fragmentation indices that describe the landscape will be altered (Neel et al. 2004).

Vegetation and Habitat Fragmentation

A second biologically based issue with fragmentation analysis is associated with the mismatch between vegetation fragmentation (what is generally indexed) and habitat fragmentation. There are times when patterns of vegetation and habitat are reasonably consistent. One example, common in the literature, occurs when wet tropical rainforest is converted to agriculture. From our human perspective, these rainforests are homogeneous. They are, however, characterized by frequent disturbances, but these tend to be associated with individual treefalls and associated regrowth (due to the human scaling associated with Fig. 3.1, studies of these disturbances are generally not referred to as "disturbance ecology," but rather as "gap dynamics"). Nevertheless, because these disturbances are small, compared with the resolution of most imagery, mature rainforest shows up as clearly definable homogeneous patches, which are easy to separate from agricultural lands. Because agricultural areas have virtually no habitat value for most rainforest species, the remnant patches of forest become habitat islands. As they grow smaller and more distant, species with large spatial requirements are lost. From a conservation standpoint, fragmentation is a problem in this situation, and habitat fragmentation can be mapped with minimal error, based on remotely sensed imagery (e.g., Arroyo-Mora et al. 2005). Even here, however, when correlating these patterns of fragmentation with species loss, it is important to separate the effects associated with fragmentation from those that would be associated with the loss of habitat area, regardless of pattern, which would also lead to the removal of many of these same species.

Northern spotted owls (*Strix occidentalis caurina*) in coastal Oregon provide a temperate-zone example of mapping and the role of habitat fragmentation. These birds are perch-drop predators, largely associated with dense mature forests (Thomas et al. 1990; Tempel et al. 2016). Forsman et al. (1984) showed that these owls' home ranges are much larger in areas where the majority of the landscape had been recently clearcut. Because northern flying squirrels (*Glaucomys sabrinus*) were the owls' primary prey in the area Forsman et al. (1984, 2004) studied, and these squirrels were also associated with dense forest (Rosenberg and Anthony 1992), the recent clearcuts had little habitat value. Spotted owls therefore needed to travel long distances between the remaining patches of older forest.

While we can find examples where mapping vegetation equates to mapping habitat, in most natural landscapes and in many human-modified landscapes, patterns of vegetation dispersion do not equate to habitat fragmentation. For example, for Rocky Mountain elk (*Cervus canadensis nelsoni*), habitat is a mixture of vegetation types. They need open areas to forage, densely forested areas to sleep in and to provide thermal refuges, wet meadows that stay green throughout the summer to put on weight, and arid areas with low snowpack to obtain forage throughout the winter. A landscape composed of a juxtaposition of these vegetation types is not fragmented, in the eyes of an elk. This is not to say that patterns of vegetation fragmentation are not important to ungulates. Habitat use by elk in forested areas is associated with edges (Thomas et al. 1979, 1988; Irwin and Peek 1983; Grover and Thompson 1986) in which areas containing high-quality forage and those with forest cover are in proximity. In open habitats, elk select locales combining high vegetation diversity with intermixed patches of shrubs and grasslands (Sawyer et al. 2007). Both patterns of habitat use appear to be maximized by a disturbance regime with spatial heterogeneity at relatively fine scales.

A study of Rocky Mountain mule deer (*Odocoileus hemionus hemionus*) found that their home-range size increased in areas with a few large patches of vegetation and was smallest in fine-grained vegetation mosaics (Kie et al. 2002). Mule deer depend on disturbance to create forage (e.g., Bergman et al. 2014), but the size and juxtaposition of dense forest and open patches are important. On the other hand, fine-grained disturbance mosaics appear to be optimal for white-tailed deer (*Odocoileus virginianus*), especially in areas where thermal cover is important (Wiemers et al. 2014). For this species, these fine-grained landscapes, which would be more highly fragmented (based on most fragmentation indices), are better (e.g., provide more resources) habitat than less fragmented areas.

Similarly, for northern spotted owls (*Strix occidentalis caurina*), the effects of block clearcuts are not uniform across their range. In its southern portions, their primary prey are dusky-footed woodrats (*Neotoma fuscipes*), which are positively associated with the brush fields that characterize recent clearcuts. Sakai and Noon (1993) showed that in brushy-phase clearcuts dominated by tanoak (*Notholithocarpus den-*

siflorus), woodrats reached densities 50–200 times as high as in old-growth forests (woodrats were absent from younger forests). These owls, being perch-drop predators, could not directly hunt in these brush fields, but they could effectively work the perimeters (Zabel et al. 1995). Thus the configuration of the landscape and the juxtaposition and patch size of the clearcuts is important to northern spotted owls in areas where woodrats are their dominant prey, but their habitat is not fragmented in the way Forsman et al. (1984) demonstrated, farther north. To reiterate, patterns of plant community fragmentation affect species' habitats, but this does not necessarily imply habitat fragmentation. Nor do increased levels of plant community fragmentation per se necessarily equate to poorer quality habitat.

In the examples we provided above, it is essential to note that species-specific definitions and delineations of landscapes will differ between the predators (e.g., owls, ungulates) and their prey (e.g., squirrels, forage). They overlap, but they will also differ in relation to each other and over time (e.g., elk might migrate, because their forage disappears during winter). The utility of research, often translated into management recommendations, will be enhanced when we avoid the human perspective of landscape and make sure to evaluate how each species we are examining moves through the environment, and how the landscape they perceive changes with them. To drive this point home, consider how elk look at their surroundings, relative to the shrews they tower over.

Like habitat fragmentation (most commonly defined by patterns of vegetation), other measures of habitat heterogeneity can take many forms. One kind of heterogeneity is temporal fragmentation, sometimes called "ecological continuity." This refers to the degree to which a particular environment, such as an old-growth forest, occupies a specific area over time (McMullin and Wiersma 2019). If an old-growth forest ecosystem is interrupted, such as with widespread forest conversion or cutting, and then allowed to regrow, many of the original species closely associated with such environments may nonetheless be lost. Thus regional and site histories are important in interpreting community composition and species occurrence.

Some research on the potential problems of ecological continuity, particularly of old-growth forests, has been done in Europe (Herold and Ulmer 2001), although the concept is still relatively new in North American land management. European researchers have discovered that vascular plants have differential adaptations to the degree of ecological continuity in the taiga of Scandinavia (Delin 1992). Some cryptogams (lichens and bryophytes, including mosses) are adversely affected by temporal disruption of old-growth forest conditions and thus can serve as indicators of ecological continuity of such environments (Tibell 1992; Selva 1994). Similar work has been done on fungi (Selva 2003; Sverdrup-Thygeson and Lindenmayer 2003). Much work remains to be done to determine sensitivity of wildlife species to ecological continuity, both within resource patches and across landscapes, although models of individual responses to habitat patch configurations might provide a useful tool (see Chapter 5).

Habitat heterogeneity also can refer to temporal variation in environmental conditions or resource availability. The time dimension is usually modeled as the independent variable in population viability analyses that assess persistence of a population in heterogeneous (fragmented) habitats. Some authors have found that temporal variation in habitat conditions can mediate whether species can coexist (Billick and Tonkel 2003; Holt and Barfield 2003). Schmidt et al. (2000) modeled populations under several spatial and temporal variations in habitat conditions related to demographic fitness. They concluded that it is important to consider the effects of both spatial heterogeneity and temporal variability when analyzing the coexistence of competing, vagile organisms. At broader scales of geographic extent and evolutionary time, Thompson (1999) suggested that the coevolution of species is an important ecological condition that continually reshapes interactions across spatial and temporal scales and, thus, helps structure communities.

Heterogeneity and fragmentation can also refer to subtle discontinuities in environmental conditions, rather than to changes in just gross vegetation structure and successional stages. One example is the horizontal separation of vegetation within a stand, such as among canopies of large trees. This kind of fragmentation has sometimes been called "within-stand patchiness" or "alpha-diversity of vegetation structure" (Kitching and Beaver 1990). In forests, this would probably adversely affect arboreal-dwelling species requiring contiguous canopy structures, such as some primates and rodents that rely on complex, 3-dimensional runways through the forest canopy. Species likely to require such conditions include lion-tailed macaques (*Macaca silenus*), a highly endangered primate of wet evergreen rainforests in southern India; red tree voles (*Arborimus longicaudus*) of western North American conifer forests; and Indian giant squirrels (*Ratufa indica*) of deciduous and moist evergreen forests in peninsular India. Other species that need dense forest canopy conditions but may be less vulnerable to forest gaps and canopy separation may include Nilgiri langurs (*Presbytis johni*) and common giant flying squirrels (*Petaurista petaurista*) in South India, and northern flying squirrels (*Glaucomys sabrinus*) in North America. Doubtless, other species can be added to these lists. The effect on and the number of species that would be included there depend on the degree of facultative or obligate use by the species.

Another poorly studied and subtle aspect of fragmentation is the vertical separation of vegetation layers, such as forest canopies and understories. The degree of heterogeneity of a vertical forest-stand structure is well known to correlate with bird species' diversity (Carrasco et al. 2019). It may also influence the use of stands and landscapes by forest-dwelling raptors that fly and forage below the canopy, such as broad-winged hawks (*Buteo platypterus*), northern goshawks (*Accipiter gentilis*), and some owls. Vegetation-layer diversity can be greatly affected by the management of vegetation, such as a silvicultural thinning of forests and the burning of woodlands and grasslands (Sullivan et al. 2002; Hayes et al. 2003; Patriquin and Barclay 2003; Suzuki and Hayes 2003). Light detection and ranging (LIDAR) holds promise as a tool to index vegetation layering, as it can record 3-dimensional attributes of forest canopies (Lefsky et al. 2002; Mao and Hou 2019).

Although fragmentation is most frequently viewed as having negative consequences for species conservation, the heterogeneity of environmental conditions can have a positive effect on some species-assemblage measures. Fahrig (2017) reviewed 118 studies summarizing 381 responses (e.g., changes in species density or diversity) by species to fragmentation. (For a complete list, see Fahrig 2017:Supplemental Appendix.) Of these, 76 percent were positive (e.g., increasing species richness) responses. Underlying mechanisms for these positive responses included increased functional connectivity, habitat diversity, positive edge effects, stability of predator-prey/host-parasitoid systems, reduced competition, spreading out of risks, and landscape complementation. The following are some rationales provided by the authors in Fahrig's (2017) review. Functional connectivity was enhanced, because smaller but more-numerous patches increase the likelihood of encountering patches that would enhance immigration and reduce emigration. Habitat diversity provides more opportunities for a greater number of species and perhaps reduces regional extinction risks. A larger amount of edge is consistent with the axiom first offered by Leopold (1933), in that it enlarges the diversity of resources to then support more species. Fragmentation can increase the persistence of predator-prey and host-parasitoid systems by providing refugia for prey or host, or by reducing the dispersal of the predator, thereby allowing the prey to stay a step. For species that might require multiple combinations of resources to meet different life requisites (e.g., the elk example provided earlier, in the introduction to this chapter), fragmentation leads to

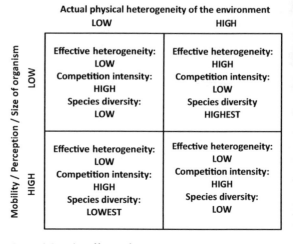

Figure 3.3   The effects of species mobility, physical heterogeneity of the environment, and competition on species diversity. Effective heterogeneity is the combination of the first 2 elements. Mobility can also be depicted as the ability of organisms to perceive their environment and may be crudely correlated with body size. (Reproduced from Huston (1994:95, Fig. 4.4), with the permission of Cambridge University Press)

complementation and easier access to the species' varying needs. Huston (1994) posited that spatial heterogeneity of the environment could interact with species competition and the mobility and size of organisms to determine species diversity (Fig. 3.3). In this model, spatial heterogeneity is less effective in preventing competitive exclusion among mobile organisms than among sessile ones. Thus species diversity is highest in environments with high physical heterogeneity, as well as with species of low mobility or small body size and low competition intensity.

Measures of Habitat Heterogeneity and Fragmentation

A wealth of mathematical indices have been offered as ways to measure habitat heterogeneity (Li and Wu 2004). These indices are typically calculated using geographic information systems (GIS) and computer software packages, such as FRAGSTATS (McGarigal and Marks 1995) or the Indian program BIOCAP (IIRS 1999; Roy and Tomar 2000). These measures derive from geometry, topology, graph theory, matrix algebra, and other fields, and they are ways to represent and analyze spatial arrangements of habitat patches over a surface.

The use of these indices often results in confusion. An index in general should be thought of as a statistical estimator of some ultimate parameter. Most users of indices of habitat heterogeneity fail to clearly identify the parameter being indexed. Claiming that fragmentation, or habitat patch pattern, or habitat heterogeneity is the parameter is inadequate. And defining the parameter in terms of the index—for example, a porosity index is an estimator of the porosity of habitat patches—is tautological.

Similar to the confusion associated with equating vegetation patterning as habitat patterning, an index of habitat heterogeneity should be expressed as an estimator of some *biological* parameter. For instance, if a wildlife population is known to be dispersal limited—that is, its population size, density, or rate of increase is circumscribed most by the death of or lack of breeding by dispersing individuals within a landscape—then one should use measures of habitat patch connectivity or permeability (Chan-McLeod

2003) as indices or statistical estimators of the ultimate parameter of (biological) dispersal success within a landscape. An example is the study by St. Clair (2003), who found that forest-dependent songbirds were more reluctant to cross rivers than roads or meadows, and that their response to such potential barriers to movement arose from the degree of forest dependency of the species. Thus a landscape with roads and meadows would be more permeable to forest songbirds than a landscape with rivers (although crossing roads carries the risk of vehicular collision), so that, following our principle above, these landscape features can serve as indices of the biological parameter of successful movement by those birds.

Such cautions taken, we can now explore a few selected indices of habitat heterogeneity (Tables 3.2 and 3.3). The simplest measures are total habitat area within a landscape and the area of individual habitat patches. A more complex index is that of patch shape, which was actually published first as an index of the shape of lakes in the physical limnology literature and later reinvented by wildlife biologists (Fried 1975; Patton 1975; Marcot and Meretsky

Table 3.3.   Some indices of habitat heterogeneity

Index	Definition	Formula	Variables
Shape	Degree of deviation from a circle	$D = \dfrac{L}{2\sqrt{\pi \cdot A}}$	L = circumference of the path A = area of the patch
Patchiness	Relative size and isolation of vegetation cover patches	$P = \dfrac{\sum\limits_{i=1}^{n} D_i}{N} \cdot 100$	N = number of boundaries between adjacent patches D_i = dissimilarity value for the ith boundary between adjacent patches
Porosity	Number or density of patches within a particular type of vegetation, regardless of patch size	$PO = \sum\limits_{i=1}^{n} C_{pi}$	n = number of cover classes among all patches C_{pi} = number of closed patches of the ith cover class
Interspersion	A count of dissimilar neighbors of a given patch and the intermixture of vegetation cover types across a landscape	$I = \dfrac{\sum\limits_{i=1}^{n} SF_i}{N}$, where $SF_i = \sum\limits_{i=1}^{n} \dfrac{Edge}{\sqrt{(Area_j \cdot \pi)}}$	$Edge$ = length of edge of the patch in both x and y directions $Area$ = area of the jth patch, formed by groups in the ith cover class

Source: McGarigal (2015)

1983). Moser et al. (2002) suggested a new index of patch shape, which served well to predict the species richness of vascular plants and bryophytes in rural landscapes. Orrock et al. (2003) found that patch shape, along with habitat corridors, influenced seed predation by invertebrates, rodents, and birds. Ohman and Lamas (2005) used a shape index to reduce edge effects and forest cover fragmentation in long-term forest planning.

More complex indices exist for calculating patchiness, porosity, and interspersion (Table 3.3). Although it is tempting to refer to all of these as indices of fragmentation, there is actually a specific fragmentation index that represents the comparative extent of different patches and is calculated as the number of patches of each cover type per unit area. As with other indices, the fragmentation index can be normalized (e.g., in a range of 0 to 10) to compare either different landscapes or a single landscape over time.

A disturbance index can be calculated from a linear combination of other landscape indices, using weighting factors for each index. The determination of weights can be based on adjacency of types of vegetation in an analysis of patch juxtaposition. Researchers should be aware that many of these indices of habitat patch pattern are highly correlated (Fig. 3.4).

There is a plethora of heterogeneity metrics that have been suggested in the literature, including indices of landscape cohesion (Opdam et al. 2003); lacunarity (McIntyre and Wiens 2000); landscape division, splitting, and effective mesh size (Jager 2000); aggregation (He et al. 2000); permeability (Stamps et al. 1987); and contagion (Parresol and McCollum 1997). (Also see indices listed by Forman and Godron 1986:188–189; Wagner and Fortin 2005.) Giles and Trani (1999) took this a step further and suggested use of a single composite index that collapses 6 factors: area, class, proportion of dominant class, number of polygons, polygon size variance, and elevational range. Although such composite indices help reduce the dimensionality and complexity of a problem (e.g., the various data reduc-

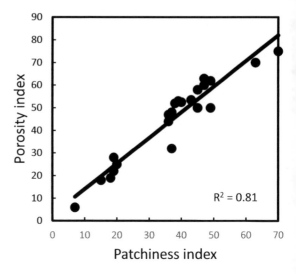

Figure 3.4 An example of how 2 indices of a habitat patch pattern can be highly correlated. (See Table 3.3 for the formulae.) The data are from a study of Asian elephant (*Elephas maximus*) habitat relationships in India (Marcot et al. 2002). The axes represent the percent of individual elephant census zones that have high levels of either patchiness or porosity. The 2 indices are highly correlated, so a person would do well with choosing the index having the greater biological meaning to represent habitat patch patterns. In this case, however, it is unclear which index fulfills this criterion for Asian elephants. The patchiness index may represent the amount of hiding cover or food resources available within a forest patch, whereas the porosity index may represent traversability among patches. Both of these factors influence elephant occurrence and density. (Reproduced from Morrison et al. (2006), with the permission of Island Press)

tion methods of principle components analysis), they may mask a biological understanding of individual parameters. Giles and Trani (1999) warned that identical values in a multifactor index can arise from an array of very different component values.

Many authors (e.g., Nams and Bourgeois 2004; Bolliger et al. 2005) have used fractal indices, which essentially measure the degree to which patch edges intrude into patch interiors. As mentioned above, due to the spatial resolution of mapped landscapes, metrics such as patch size and edge-to-area ratios are extremely sensitive to changes in scale. Fractal metrics were designed to combat this problem. Observers noticed that the lengths of coastlines increased as the resolution increased, which led to a fundamental measurement problem: how long was the coastline, actually? It was found, however, that if you zoomed in on a coastline and re-measured it, the length of coastline grew by a fixed amount for a given increase in resolution. This fixed ratio is the fractal dimension of the coast (Mandelbrot 1975). Fractals were found to be useful in multiple fields where many objects had self-similar properties when observed at different scales. For example, fractal indices have been used to describe the foraging and movement paths of reindeer (Marell et al. 2002) and the geometric complexity of mole-rat burrows (Le Comber et al. 2002), as well as to simulate landscapes (Hargrove et al. 2002).

In addition, indices of habitat heterogeneity have been based on graph theory (also called "loop analysis"). In this approach, habitat patches are usually represented as nodes or points, and the adjacent edges between patches as arcs or lines between the points. Various patch pattern indices borrow from graph theory measures of connection complexity (Marcot and Chinn 1982; Cantwell and Forman 1993; Urban and Keitt 2001).

Thus far we have discussed indices and measures of habitat heterogeneity across space. Other tools are useful for indexing, measuring, and modeling habitat heterogeneity over time. Early models of ecological succession were based on a Markovian analysis (Horn 1975), which is a statistical technique that represents transition probabilities among successive states of a system. More-general transition matrix and Markovian approaches were later used to model the sensitivity of successional changes to management activities, including harvesting, grazing, fire, and other manipulations (Hann et al. 1998; Rojo and Orois 2005).

Connectivity

At its simplest, connectivity is the converse of fragmentation. Of note, however, the analysis of connectivity is quite distinct from fragmentation analysis, as it uses very different algorithms. It also has its own issues, which are associated with both the presentation of landscape features and behavioral assumptions that generally are not closely linked to biology. Unfortunately, but probably not surprisingly, connectivity analyses are fraught with many of the same problems as fragmentation analyses, and they add other unique problems of their own.

Generically, landscapes are likely to be connected if the proportion of habitat on the landscape exceeds the percolation threshold (Stauffer and Aharony 2018). A non-technical definition of percolation threshold is the density of barrier objects, for which (assuming a very large field of such objects) there is always a way through. Percolation theory generally assumes some sort of lattice of potential habitat areas (representing a landscape as a raster is one of many possible geometries) from which areas are removed randomly. Given this, and depending on the configuration of the lattice, you need between 50 and 80 percent of the landscape to be habitat, in order to ensure that the random removal of habitat blocks will not lead to barriers (areas where there are no connecting habitat blocks). While this approach is highly stylized—in the real world, habitat does not occur and is not removed in equal-sized blocks through random processes—the simple take-home is that if you have well over 50 percent of the landscape as habitat, habitat connectivity most likely will not be problematic. Similarly, if you have under 50 percent of the landscape as habitat, you probably need to actively manage it to make sure that habitat connectivity is maintained. As a reminder, we are discussing landscape within the context of a known (the best situation) or presumed (at least you thought about it) biological population.

How much connectivity is needed? This depends entirely on what population process that connectivity is serving. Connectivity can link habitat elements in daily use, such as access to a water source. It may allow seasonal access to resources, such as winter range for ungulates. At population scales, which are the focus of much of this book, connectivity can promote the ebb and flow of a population across the landscape as that population increases and decreases, a process conceptualized as an ideal free distribution (Fretwell and Lucas 1970; Fretwell 1972). Fretwell and Lucas (1970) conceived of a population in which all organisms inhabited locations promoting high survival and productivity, given the population size. When the population was growing, areas would be excluded, since they were already occupied, and organisms would be forced to inhabit sites with lesser potential quality. As populations shrank, areas with a higher probability of promoting survival and productivity would become available, and organisms would shift, in order to fully occupy those habitats. Clearly, these processes cannot occur without significant levels of habitat connectivity. In more highly clumped habitats, where they form fairly large discrete islands, connectivity may serve to add individuals to support a distant subpopulation (demographic rescue) or genetic material (genetic rescue). In both cases, connectivity allows a group of subpopulations to potentially form a larger metapopulation (see also Chapter 1).

Levins (1969, 1970) first modeled metapopulation dynamics. His model, like the concepts of percolation and ideal free distribution, was highly stylized, with subpopulations that were all of the same size and had the same connectivity to other subpopulations. In this model, subpopulations became extinct at a fixed rate, and habitat islands were recolonized with organisms from other subpopulations, also at a fixed rate. Additionally, it was a dynamic equilibrium model. Its primary finding, however, which is somewhat counterintuitive, has enormous heuristic value. Levins (1969, 1970) discovered that the equilibrium proportion of subpopulations that would occupy habitat islands was equal to 1 − (extinction rate / colonization rate). That is, when the colonization rate is less than or equal to the extinction rate, the metapopulation is at risk of becoming extinct. This is noteworthy, because one might intuitively think that if colonization equaled extinction,

at least some proportion of the subpopulations would occupy the islands.

What Levins's (1969, 1970) models indicated was that if you have a system of separate and similar-sized islands, and if you want most of your islands to be occupied, colonization must greatly exceed the rates of local extirpation. Such a system will be highly dependent on connectivity, particularly if the subpopulation sizes are low, say less than 50 (Lande and Barrowclough 1987). Hanski and Gilpin (1991) argued that a far more common arrangement is to have a few large islands with many smaller islands, termed a "mainland-island metapopulation." This structure is much more globally stable than the equal-sized patch model that Levins (1969, 1970) envisioned, because the large blocks have very low extinction rates, due to their substantial population sizes, and hence are not as dependent on colonization. The effects of connectivity on populations have received a great deal of attention, both in the literature and in the practical worlds of population management and the design of reserves, although this is not the largest scale at which connectivity is important.

As with fragmentation, connectivity has both positive and negative consequences. For example, while it is true that connectivity is necessary for metapopulation persistence, a correlation between subpopulations increases the global extinction rate of a metapopulation and reduces the positive benefits associated with increasing the number of reserves (Goodman 1987; Quinn and Hastings 1987; Gilpin 1990; Hanski and Gilpin 1991). Correlations can occur because of environmental events, such as hurricanes or droughts, or they can be biological in origin, since diseases and exotic competitors and predators travel along lines of connectivity. These dual attributes of fragmentation are perhaps most clearly seen at the largest geographic scales. Isolation leads to speciation, which increases the richness and biodiversity of the planet; it also, however, leads to extinctions (Hubbell 2001). Connectivity, or the lack thereof, during and immediately after the Pleistocene had profound effects on the biogeography of both North American and Eurasian biotas (for a discussion, see Pielou 2008).

Connectivity is also critical at biogeographic and evolutionary scales, with negative consequences occurring at the related temporal domains. Failure to obtain a daily requirement, such as water, will lead to negative consequences at a time scale ranging from days to months. Failure to obtain necessary seasonal resources will lead to negative consequences at time scales of a few years. Failure to provide demographic and genetic rescues will have negative consequences at time scales of generations; for many species, this could range from decades to centuries.

Similar to the evaluation of fragmentation, a group of algorithms (see Wade et al. 2015 for a complete summary) are used to assess landscape connectivity. Whereas fragmentation analysis involves assessments of the pattern, size, shape, and juxtaposition of patches, connectivity analysis primarily uses routing algorithms to find likely pathways across a landscape. These routing algorithms are generally point-to-point. That is, an organism is assumed to start at a particular location in the landscape and seeks to travel efficiently to some other specific location. Such algorithms ask, if an organism started at some location and wanted to get to another location and had no ancillary goals or constraints, what route would it take?

Measures of Connectivity

One common connectivity algorithm is the least-cost path, or LCP (e.g., Cushman et al. 2006, 2009). LCP algorithms optimize potential travel routes globally—that is, the algorithm may denote routes that are locally suboptimal to achieve an optimization of connections across the entire landscape. If you have used a digital driving assistant in your vehicle, you will be familiar with these algorithms. Full global knowledge of the costs associated with all possible paths is critical in using such algorithms. Although your digital assistant may have this knowledge, a dispersing organism generally would not. While this has served as a criticism, the primary critique of LCP algorithms in the literature is that the path is only one pixel wide

and, therefore, is both insensitive to the environment around the LCP and overly sensitive to pixel-level attributes. (See Wade et al. 2015:Table 2 for a list of strengths and weaknesses associated with connectivity algorithms). These problems are trivial to fix. If you want pixel attributes to reflect the surrounding landscape (e.g., you want a pixel for high-quality habitat to be more valued if it is surrounded by other high-quality habitat pixels), you can adjust their values to reflect this. Most types of GIS have neighborhood functions (e.g., ESRI's ArcGIS focal mean) that can accomplish this.

To relax the single pixel–wide path that LCP algorithms generate, several methods have been employed; collectively, these are referred to as least-cost corridors (LCC). A simple approach is to buffer the LCP—that is, make it wider than one pixel (e.g., Cushman et al. 2009). This approach has to be viewed as primarily useful for display purposes. For example, in a model with many source and destination locations, areas where many paths converge will be effectively highlighted by placing kernels along each path and summing the kernel values. A more biologically meaningful approach is to retain both the optimal and all of the nearly optimal paths as possible travel routes. As these paths generally fall among similar routes (the nearly optimal routes usually only vary at a few pixels), they identify broader areas adjacent to the optimal path (e.g., Pullinger and Johnson 2010).

Another common algorithm envisions the landscape as a web of resistors (each pixel is a resistor at the junction of four wires) and organism connectivity as electrical flow across the landscape (McRae and Beier 2007; McRae et al. 2008). In this approach, source and destination locations are seen as anodes and cathodes in an electrical circuit. Because circuit flows follows Ohm's Law, wider areas with low resistance receive more current flow than narrow areas, generally seen as a desirable property for conservation planning. Additionally, current flows, to some degree, through all available paths, highlighting alternative routes. Though optimal LCP or

LCC paths are generally consistent with the paths associated with the dominant current flow (see Carroll et al. 2012:Fig. 3), electrical flows highlight a much broader range of alternative paths. Once generated, connectivity maps can be used directly, to identify areas of interest, or they can be evaluated as networks (Urban and Keitt 2001). Viewing lines of connectivity as a network allows the application of a suite of network statistics, such as centrality (Upadhyay et al. 2017), allowing the relative merits of particular connectivity routes to be evaluated thoroughly and quantitatively.

Unlike analyses of fragmentation, connectivity has no explicit links to classified landscapes—the algorithms can operate equally well in continuous gradient space—but classified landscapes are often used in the process. If classified maps serve as the basis for connectivity modeling, then resulting connectivity analyses are subject to the same issues regarding rules for subjective typing and the loss of subpixel detail discussed above. In addition, connectivity analyses contain a suite of unique and generally untested premises that have tenuous links to biology. Two of these were mentioned above. To reiterate, the first is that in an LCP, organisms know the details of all possible routes between 2 locations and choose the very best. The second is that organisms flow across the landscape in a manner similar to the electron flow in a circuit. Both of these concepts appear unlikely to be correct in any strict sense.

Perhaps more importantly, all point-to-point algorithms assume that organisms move with a strict goal orientation. When travelers use a digital assistant to find the best route to a coffee shop (using an LCP algorithm), that is all they want to do. Further, they know that a coffee shop exists somewhere, and their sole goal is to find one, fast. For this reason, an LCP works well for the morning caffeine seeker. Nevertheless, without a digital assistant armed with full knowledge of the road network and with near–real time information concerning traffic flows, accidents, and construction projects, there is no way that the caffeine seekers could, on their own, hope to approximate the path that the assistant indicates—and the travelers would probably still know more about the probable relationship between the road network and coffee shops than would a dispersing organism about the availability of distant and unseen resources. Consider a dispersing subadult leaving its natal population in search of a new home and a mate. It has no idea what resources are available, where they might be, and how difficult they might be to access. Why would we expect these dispersing organisms to take an LCP or be drawn to a distant location like electrons to a cathode?

Other problems lie in the transformation of a landscape into a surface for connectivity modeling. Before a connectivity algorithm can be used, each pixel needs to be assigned a resistance or a cost. The meaning of cost is biologically vague, but it pertains to the difficulty of an organism crossing that pixel. By far the most common assumption in the literature is that dispersing organisms find habitat easier to cross than non-habitat and, generally, that the best habitat is the easiest to cross (see Wade et al. 2015 for results from a recent literature search). This assumption is clearly tenuous, and in practice it is even more so, because "best" is only vaguely defined. In general, best equates to areas where the organism is most likely to be observed, or to the types of places where it spent the most time in a resource-selection study (Manly et al. 2002). As we have demonstrated in Chapter 2, this assumption is itself highly problematic.

Using the habitat quality paradigm and being content to define habitat based on occurrence data, connectivity models for a species are easy to make. All that is required is (1) obtain a list of occurrences; (2) run a program such as MaxEnt (Phillips and Dudík 2008; Elith et al. 2011), which is remarkably user friendly, to correlate those occurrences with landscape variables; (3) invert the resulting surface; (4) scatter source and destination nodes across the better-quality habitat; and (5) run a program like Circuitscape (McRae and Shah 2009), which is also very user friendly, to create the connectivity paths.

With this approach, one can create connectivity maps across large spatial extents during a long afternoon, where most of the time is spent waiting for the algorithms to work. Whether the resulting products are meaningful is another question entirely (see Wade et al. 2015). Again, we call for interpreting and articulating the results of such analyses as indices of some species-specific biological response.

The discussion above highlights a major problem with connectivity models: linking the presumed movements to actual biological animal movements. Two approaches have been used, both of which have merits and shortcomings. The first is the use of genetic relatedness as a proxy for movement. This method requires genetic contact; ergo, if 2 contiguous areas are highly related, they are connected through direct contact. Further, the contact needs to be of a meaningful sort. That is, if an organism travels to a distant site and does not interact with the population, its movement has less consequence than if that organism enters the population, successfully breeds, and leaves its genes behind in the form of viable offspring. Pairwise patterns of genetic relatedness can be correlated with connectivity models that are based on a variety of hypotheses, and the connectivity model that produces the highest correlation with the genetic patterns is chosen as being the most likely to represent actual pathways traveled (e.g., Cushman et al. 2006).

In our view, while this is preferable to guessing that habitat equates to movement, genetic relatedness is an indirect index of movement and has been criticized. One of the main critiques is that genetic relatedness is backward looking and may equate to past landscapes that no longer exist. For example, several generations must pass before a new landscape obstruction is detected through genetic analyses (Landguth et al. 2010). Additionally, inferential data may be insensitive to certain types of movements that direct movement data would detect. Seasonal migrations or habitat patch utilization may be entirely absent from relatedness patterns revealed by genetic data (Spear et al. 2010).

An alternative is to collect path data directly, which has now been simplified for larger species through GPS technologies. One promising tool for transforming path data into rules that can be applied to a connectivity model is the application of state-space models, a class of time-series models that predict a system's future state, based on a probabilistically derived process model (Horne et al. 2007; Patterson et al. 2008). State-space models estimate the probabilities of a particular state (e.g., location), model variables (mean speed and turn-angles), and subsequently incorporate these parameters into a flexible future forecasting model. The combination of these models with fine-scale satellite telemetry data is rapidly improving our understanding of animal dispersal and movement (Horne et al. 2007). For example, Vergara et al. (2013) used state-space models and translocation experiments with austral thrushes (*Turdus falcklandii*) to show that the use of riparian strips for connections between habitat patches is highly influenced by the surrounding landscape.

The Concept of the Corridor

Given the dissection of wildlands by urban, agricultural, and other anthropogenic areas, managers often seek to minimize the perceived negative effects of fragmentation and enhance the positive effects of connectivity by either preserving or generating corridors. Given 2 blocks of vegetation separated by what is viewed as an impenetrable barrier, a common approach is to create a pathway through the impenetrable landscape to provide connectivity. These structures have commonly been dubbed corridors, and they have a long and controversial history (e.g., Simberloff et al. 1992; Rosenberg et al. 1997).

To be effective, corridors need to be found by the moving organisms, provide habitat for at least safe passage through them, and lead to somewhere useful (e.g., Haddad et al. 2003). For short corridors, such as highway underpasses or overpasses, fencing is generally needed to funnel organisms into the corridor and keep them in it, once entered. Because both finding the corridor and achieving safe passage through it can be engineered, these corridors tend to be effective, particularly if they connect to someplace desirable, such as access to ungulate winter range (e.g., Plumb et al. 2003). The efficacy of longer, more amorphous, unfenced corridors is less clear and is dependent on the nature of the matrix in which they are embedded (Vergara 2011).

One critical property of corridors that is frequently understressed is whether the target organisms can live and reproduce within the corridor or can only employ it as a travel conduit. Many successful uses of corridors associated with the spread of exotic species are due to the ability of these species to live in a corridor within an otherwise hostile matrix. If a corridor can hold a population, then spread can occur in generational time, such as allowing eastern North American species to colonize the West. One example is barred owls (*Strix varia*) (Livezey 2009). The efficacy of corridors used exclusively as travel corridors is much more speculative.

Ultimately, there is no best corridor configuration for all species. Rather, the design and establishment of corridors must be based on an analysis of species-specific habitat and resource requirements, their dispersal ability, and their overall behavioral responses to a patchy environment. We are not suggesting that every species needs a different type of corridor. Rather, each species must be evaluated to determine the likelihood of its using a corridor, including a use relative to its movements outside of the corridor. If corridor use is unlikely, then alternative means of ensuring that a species occupies a specific location can be considered, such as a periodic augmentation of individuals. (For a detailed discussion of corridor development and management, see Morrison 2009:127–134.)

Disturbance Ecology

In landscapes, much of the observed heterogeneity is due to disturbance processes. The study of disturbance ecology focuses on dynamics of habitats in landscapes. It addresses many topics that range widely across

scales of space and time (Frelich 2002), including soil dynamics, fire ecology, vegetation succession, meteorology, climatology, and paleoclimatology. Similar to landscape ecology, however, most manuscripts that claim to study disturbance are focused on a range of disturbances that bracket spatial scales we associate with disturbances and temporal scales that are relative to human lifespans. To best understand their effects on animals, disturbances should be depicted according to their frequency, intensity, duration, location, and geographic extent (Engstrom et al. 1999). In this way, disturbances to habitats can be incorporated into spatially explicit models of population demography to project effects on species distribution and viability (Boughton and Malvadkar 2002; Dunning et al. 1995).

Four Types of Disturbances

Given human-oriented scaling constraints (e.g., here we are not considering the spread of continental ice to be a disturbance, although it has had profound effects on species distributions), we can broadly classify disturbances into 4 categories, based on their intensity and geographic extent (Fig. 3.5).

Type I Disturbances

Type I disturbances are major environmental catastrophes that are relatively short term, intense, and affect large areas. They include volcanoes, major fires, floods, and hurricanes (labeled typhoons in the northwestern Pacific Ocean, and cyclones in the southern Pacific and Indian Oceans). Examples of Type I disturbances that have been studied in North America include the 1980 eruption of Mount St. Helens (Lawrence and Ripple 2000), the 1988 Yellowstone National Park fire (Turner et al. 2003), Hurricane Matthew affecting coastal Georgia in 2007 (Gibson et al. 2018), and the 1993 Mississippi River floods (Custer et al. 1996; Miller and Nudds 1996). Each kind of Type I disturbance can bring very different changes to environments and species composition, and they deserve individual attention in hab-

Geographic area affected		
	Widespread (>1000 ha)	Local (1-1000 ha)
Degree of disturbance — high	Type I *Major environmental catastrophe* (volcanoes, major fires, hurricanes)	Type II *Local environmental disturbance* (wind, ice storms, insects, disease)
Degree of disturbance — low	Type III *Chronic or systematic change over a wide area* (predators, competition, forestry)	Type IV *Minor environmental change* (local fires, developments)

Figure 3.5   Four types of disturbances, shown by their degree, or intensity, and the geographic area affected. (Reproduced from Morrison et al. (2006), with the permission of Island Press)

itat planning that accounts for such infrequent events. Some Type I disturbance events occur in multiple cycles, such as fire regimes or floods with various repeat frequencies.

Foster et al. (1998) studied 5 examples of Type I disturbances (which they called LIDs, or large infrequent disturbances), including fire, hurricanes, tornadoes, volcanic eruptions, and floods. They concluded that the resulting landscape patterns were controlled by the type of disturbance, topography, and vegetation. They found that different kinds of LIDs produced dissimilar edge effects and varying extents of enduring legacies that could influence ecosystems for decades or centuries.

Hurricanes have a special place among Type 1 events, due to both their size and frequency. Disturbance areas associated with hurricanes are large. Moreover, between 1970 and 2010, 637 hurricanes made landfall worldwide (Weinkle et al. 2012), making their occurrence common. Hurricanes can have major effects on wildlife habitats and populations, and, in some cases, habitat refugia can serve as important protection zones for populations. Hurricanes can affect species and resources differentially. For example, Pierson et al. (1996) found that 2 severe cyclonic storms on Samoa caused a more severe population decline in the more common, widely distributed species of flying fox (*Pteropus tonganus*) than in the endemic species (*P. samoensis*). They at-

tributed this difference to the greater susceptibility of *P. tonganus* to hunting mortality in villages, as this species appeared in those areas to forage on flowers and fruits after the storms. Their greater susceptibility was also due to the larger proportion of foliage in the diet of *P. samoensis*, whereas *P. tonganus* is more highly frugivorous and, overall, the post-storm density of flowers and fruits was depressed. The endemic species was protected from storm effects in rainforest refugia, particularly in areas of high topographic relief (e.g., volcanic cones and steep valleys) were less subject to wind damage, whereas the more widespread species occurred in the less protected village environments that were more severely damaged.

Using clipping experiments to simulate pruning effects of Hurricane Lili, which in 1996 directly hit the island of Great Exuma in the Bahamas, Spiller and Agrawal (2003) concluded that hurricanes can trim shrubs, resulting in new growth that entices up to a 68 percent increase in herbivory by arthropods. They argued that such increased sprouting and herbivory following natural or anthropogenic disturbances have wide implications for understanding how storms affect habitats for organisms.

One rather well-studied storm was Hurricane Hugo, which can serve here as an example of major effects of Type I storm disturbances on wildlife and habitat. Hugo hit South Carolina on 21 September 1989, with sustained winds of 217 km/hr, gusts of 282 km/hr, and a storm surge of 5.8+ m. Damage to forests by Hugo was greater than that from Hurricane Camille, the eruption of Mount St. Helens, and the 1988 Yellowstone Fires combined. Hugo damaged forests on over 17,800 sq km, with this being greatest in coastal Francis Marion National Forest (NF), where about 75 percent of commercially marketable pine trees fell during the storm (Ehinger 1991). Many forests in South Carolina that were hit by Hugo had held some of the densest colonies of endangered red-cockaded woodpeckers (*Picoides borealis*) found on the East Coast of the United States. Prior to Hugo, Francis Marion NF supported about 25 percent of all southeastern red-cockaded woodpeckers, including one of the world's largest populations of approximately 500 breeding pairs (Walters 1991). On Francis Marion NF, most of the damaged trees were mature pines favored by these woodpeckers. Hugo reduced the woodpecker populations there by about 60 percent and destroyed 87 percent of the known active cavity trees (Walters 1991). Recovery of forest conditions for red-cockaded woodpeckers could take 75 years, and demographic consequences to woodpecker populations there will probably be long lasting (Hooper et al. 1990).

Hugo also passed directly over Puerto Rico, causing heavy damage, including the breakage and blowdown of many trees and an almost complete defoliation of the remaining trees (Walker 1991). Initial censuses showed that only 23 of the original colony of 46 wild Puerto Rican parrots (*Amazona vittata*), a highly endangered species, remained. Analyses by Beissinger et al. (2008) suggested that hurricanes are the major cause of demographic bottlenecks of this species, limiting its population growth. A captive flock of 53 parrots was unhurt, however, which is a prime example of the value of captive breeding for at least some highly endangered species subject to periodic major disturbance events in the wild. After the storm, many predators were observed in areas of heavy treefall. Together with destruction of nest structures caused by Hugo, predation can seriously depress some populations. Nevertheless, the parrot population has since recovered to levels approaching that of pre-Hugo, thanks to the surviving and actively reproducing birds, the availability of improved natural nest cavities, and the effectiveness of an enhanced nest management program. Since Hugo, some Puerto Rican parrots have nested in lower elevations and used cavities in tree species never before reported since parrot observations began in 1968 (Vilella and Garcia 1995).

The effects of Hurricane Hugo on other forest bird species in Puerto Rico were studied by Wunderle (1995), who reported that canopy dwellers shifted to foraging in forest understories and openings, and bird assemblages became far less distinct by microhabitat conditions than they had before the hurricane. Such habitat displacements of birds also included movement of frugivorous birds into preexisting gaps and an invasion by forest-edge or shrubby second-growth species into the forest. Wunderle (1995) speculated that it may take many years for resources and structures in forest understory and gaps, along with associated unique bird assemblages (as observed prior to the storm), to again become distinct.

Type II Disturbances

Type II disturbances include locally intense environmental changes from events, such as wind storms, ice storms, and local outbreaks of defoliating insects. Type II wind events can include microbursts and tornados. Forest canopy gaps are sometimes called "microserules." They undergo local succession by plant species and are important contributors to the overall vertical forest-stand structure and species composition (Moeur 1997; Degen et al. 2005). For example, Lawton and Putz (1988) reported that wind-formed canopy gaps of tropical montane elfin forests in Monteverde, Costa Rica, tended to be small, occur frequently, and annually cover small proportions ($\leq 1\%$) of the overall forest cover. These gaps promoted the growth of shade-intolerant plant species, tree saplings on nurse logs and mineral soil disturbed by uprooted trees, and plants coming from soil seed banks more than from seeds dispersed into the gap. The researchers also found that many saplings in gaps originated from epiphytes in the crowns of the trees that had fallen. In temperate forests of northern Minnesota, Webb (1989) observed that tree damage from thunderstorm winds (25–35 m/s) was related to tree size, species, wood strength of those species, and incidence of species-specific fungal pathogens. He found much of the within-stand variations resulting from wind-caused mortality and subsequent plant development in the canopy gap were due to differences in shade tolerance, initial forest structure, gap size, and wind-firmness of shade-tolerant understory plants.

The size of forest canopy gaps also influences nitrogen-cycling dynamics, as reported by Parsons et al. (1994) in lodgepole pine (*Pinus contorta*) forests in southwestern Wyoming. In these forests, removal of 15–30 tree clusters represented a threshold above which significant losses of nitrogen available to the groundwater occurred.

While the general discussion of disturbance to this point has been associated with the interface between abiotic factors, such as weather and climate, and the resulting biological changes, disturbances also include events that are biological in origin, those resulting from the population dynamics of common or invasive insects. Outbreaks of defoliating insects, such as spruce budworms (*Choristoneura fumiferana*), therefore can be considered another Type II disturbance that can directly or indirectly cause changes, ranging from local to extensive, in forest structure and composition (Alfaro et al. 2001). Because these disturbances occur as ecological processes, complex species interactions often determine the control of these population outbreaks. For example, in spruce-fir forests in northern New Hampshire and western Maine, Crawford and Jennings (1989) reported that the entire bird community there showed significant functional responses (greater foraging) to increasing budworm density, whereas only 2 species, Canada warblers (*Wilsonia pusilla*) and golden-crowned kinglets (*Regulus satrapa*), also showed numerical responses (increased reproduction). The researchers

concluded that insectivorous birds are capable of dampening the amplitude of budworm infestations when habitats are suitable for supporting adequate bird populations. This corroborates other similar findings, such as that suggested for woodpeckers foraging on bark beetles in northern Colorado (Koplin 1969).

Williams and Liebhold (2000) studied the degree of correlation among spruce budworm outbreaks in eastern North America during 1945–1988 and found that spatial synchrony of outbreaks decreased with distance between local budworm populations, and it approached zero near 2,000 km. They also found geographically blocked clusters of outbreaks along an east–west axis. They observed that synchrony depended in part on temperature and precipitation regimes, as well as the dispersal rates of and distances for budworms, and they modeled these findings in a simulation. Their results suggest that defoliator dynamics are complex and depend on multiple variables.

Like pandemic levels of native pathogens (e.g., root rot) and defoliators (e.g., budworms), other Type II disturbances—especially relatively intense fires in grasslands, shrublands, and forests—often tend to leave spotty footprints, rather than entirely denuding large areas of vegetation (Harrison et al. 2003). In this way, similar to smaller canopy gaps created by tree windfalls, fires often produce complex vegetation patterns.

As patches of vegetation change over time, so do responses by animals as they use these patches for breeding, foraging, finding a refuge from predators, resting, and dispersing. Some wildlife species probably evolved in concert with native disturbance regimes and take optimal advantage of resources distributed through space and over time in shifting patches. Thus the further that human activities alter native patch disturbance dynamics, the greater the discontinuity may be with the evolved habitat selection behaviors of certain species. In some cases, even if a suitable environment is present but is greatly altered in its patch distribution pattern and temporal occurrence, an associated wildlife species may be excluded (Ellner and Fussmann 2003; Summerville and Crist 2003). This aspect of coevolution with native disturbance regimes, native resource patch dynamics, habitat selection behaviors of wildlife, and effects of changes in native disturbance regimes on wildlife viability and fitness is little studied and deserves research attention.

Type III Disturbances

Type III disturbances are chronic or systematic changes over wide areas. They include the slow alteration of native landscapes by human habitations, ecological succession, and long-term climatic changes. Wildlife relationships to Type III disturbances include variations in species abundance and in ecosystems, due to changes in regional climates (Weltzin et al. 2003). Shifts in climate can greatly alter the distribution of vegetation and wildlife over broad areas. They can also result in the rise of local sea levels, affecting the persistence of and changes in coastal marshes and their biota (Taillie and Moorman 2019).

Even climatic fluctuations lasting on the order of only a few years can greatly affect productivity of some wildlife populations, and alterations in large-scale patterns of wind and temperature can have ramifications on many systems worldwide. One example is the effect of El Niño warm water cycles in the eastern Pacific Ocean. El Niño patterns are due to a decrease in the velocity of the easterly subtropical winds in the Pacific Ocean. The reduced speeds of these winds, in turn, decreases the velocity of currents that pull cold water from the southern coast of Chile north and from the Gulf of Alaska south, producing large upwelling zones of cold, mineral rich waters. With less upwelling, tropical temperatures across the Pacific become warmer, and the waters become less productive. A large El Niño event interrupted the 1982–1983 breeding chronology and reproductive attempts of seabird communities on Christmas Island in the central Pacific Ocean (Schreiber and Schreiber 1984). As a result of the distant upwelling effect on food resources, the entire seabird community failed reproductively and temporarily disappeared from this equatorial atoll.

Grant et al. (2000) studied the effect of El Niño on the productivity of 2 species of Darwin's finches on the Galápagos Islands. They found no simple correlation between clutch productivity and El Niño–induced increases in rainfall and temperature. Rather, clutch productivity varied greatly among El Niño years, in response to prior-year weather conditions and the length of time since the last El Niño event. The authors concluded that wildlife relationships to such disturbances can be understood only in a broad temporal context. The effects of El Niño events may also extend to non-marine species far from the ocean (Stapp et al. 2004). For example, El Niño events and their converse, those of La Niña (in which an unusual amount of cold-water upwelling occurs), profoundly affect patterns of precipitation and are therefore associated with fire activities in North America, eastern Asia, the Amazon, and Australia (Simard et al. 1985; Laurance and Williamson 2001; Metlen et al. 2018).

Type IV Disturbances

Type IV disturbances include minor and local environmental changes. These include low-intensity local events, such as spot fires, low-density rural developments along the edges of natural landscapes, and the gap dynamics of vegetation canopies (Acevedo et al. 1995). In particular, small gaps in vegetation can be caused by natural plant death or by biotic mortality agents (e.g., insect defoliators, plant pathogens, and plant diseases), as well as by fire and by weather conditions such as ice or windstorms. Individual treefalls in forests serve to open up canopies, change local microclimates at the forest floor, and afford a foothold for sun-tolerant plants (Schnitzer and Carson 2001). Treefalls uproot soil masses and redistribute litter, duff, and upper soil layers, along with their associated microfauna and microflora. They can also directly affect the abundance and distribution of vertebrate wildlife species (Fuller 2000; Greenberg 2002).

Across watersheds and landscapes, disturbances affecting vegetation patch structure and composition tend to alter such ecosystem processes as surface water discharge, nutrient runoff, organic matter input to soils, net productivity, and microclimates. These changes, in turn, can influence species composition that is associated with the soil, ground surface, plant canopy, and other substrates. Changes in vegetation patches can alter energy balances of individual organisms, such as by changing the food content or values in foraging patches, thereby tipping the balance of foraging efficiency and affecting the successful reproduction and fitness of organisms.

Managing native vegetation conditions in landscapes that are subject to relatively frequent, intensive Type II and Type IV disturbances, such as stand-replacing fires, and those in which native vegetation occurs only in small isolated patches, can be a great challenge. In such cases, there may not be a sufficient area or number of patches of native vegetation to provide enough resilience in the face of intense disturbances, so significant sums and much effort must be expended to prevent changes. In some circumstances, such as the fire-prone interior west of the United States, disturbances will inevitably occur, regardless of—or, in this case, because of—efforts to thwart them (Everett et al. 1994).

Factors Affecting Disturbances

The shape and magnitude of disturbances are frequently altered, due to interactions between human activities and ecosystems. One of the best examples of this is the Dust Bowl, which occurred in the southeastern Great Plains during the 1930s. Droughts had happened periodically in this area, and the native ecosystems were well adapted to them, based around deep-rooted grasses that could withstand extended dry periods. Probably the drought that caused the Dust Bowl would have led to modest levels of ecological disturbance had the native ecosystems been intact. Large areas, however, were plowed and planted with annual species, such as wheat, leav-ing huge areas of soil open to the wind. As a result, the soil blew into lethal storms that led not only to the famous diaspora of many local residents, but also had profoundly damaging effects on the local ecosystems.

Another example of how interactions between human activities and ecosystems can affect disturbance regions can be found in the effects of fire suppression, which, in western North America, was coupled with an unusually wet period during the middle of the twentieth century. Decades of fire suppression, with the resulting build up of fuels and the high-grade, selective logging of large old-growth trees, coincided with improved conditions for forest growth and regeneration. As a result, forests in the southern Sierra Nevada have become four times denser than they were historically (Collins et al. 2017). A drought recently occurred throughout this same area. Subsequently, across elevations from 914 m to 2134 m (most of which could be considered forested in the southern Sierra Nevada), about half (48.9%) of all the trees died between 2014 and 2017. Nearly all (89.6%) of the ponderosa pines (*Pinus ponderosa*) were killed (Fettig et al. 2019). Similar to the Dust Bowl, human actions exacerbated the form and magnitude of a climate-generated disturbance.

Historically, many disturbances could be viewed as being part of stable and continuous stochastic processes. Due to this stability, species adapted to the presence of these disturbances and, in some cases, became dependent on their occurrence. Today this often not the case. Disturbance processes interact with human-caused landscape change, directional shifts in climate, and an ever-increasing number of invasive species to produce increasingly novel landscape conditions and species assemblages. One cannot take it for granted that modern patterns of disturbances and their effects on ecological responses are simply part of a benign process that has historically occurred, and that wildlife management will be best served by allowing disturbances to occur. Increasingly, historical antecedents will not provide guidance (Millar et al. 2007). We will need to formally analyze disturbance effects and link these to desired goals in a proactive manner. This approach will require a very precise formulation of goals and clear-eyed assessments of relationships among the effects of human activities on disturbance regimes, as well as between observed changes in landscape patterns and these goals.

Management Lessons

What management lessons can be drawn from this brief review of the major categories of disturbances and the dynamics of resource patches in landscapes? First, in all but the simplest ecosystems, the dynamics of vegetation and environmental factors consist of a complex medley of changes that occur on multiple schedules. Landscapes must be assessed relative to the species (or assemblage of species, or other levels of biological organization) of interest to determine which disturbance types occur, the local site histories, and the likely vegetation responses to any disturbance regimes caused or altered by management activities. For example, fire behavior can be influenced by historical factors, such as recent fire occurrences in the area; by proximate factors, such as weather and topography; by vegetation factors, such as canopy gap openings (which, in turn, are influenced by the vegetation's resistance to strong wind, its orientation to prevailing winds, and other elements); and by many additional ones. Fire, like other disturbance regimes, then influences the likelihood of further changes, such as succession, and can greatly alter the suitability and stability of the environment for wildlife.

Second, animals can play a major role in affecting both disturbance regimes and how habitats respond to disturbances, such as through predation or transportation (phoresis) of disturbance agents (e.g., forest pathogens). Herbivory has an influence on vegetation that is otherwise susceptible to disturbances. Such ecological functions of species often are not, but should be, considered in management plans that alter or introduce disturbances. Third, management

may wish to more fully study how activities change native disturbance regimes and how wildlife may respond behaviorally (functionally) and demographically (numerically) to such changes.

Fourth, the specific future responses of most ecosystems that incur disturbances at multiple scales of space, intensity, and time are not very predictable. Rather, what can be better predicted, at least as probabilities or as frequencies, are the disturbance regimes themselves. In this sense, management can then identify a set of desired future dynamics, perhaps to reconstruct or mimic native disturbance regimes in which some species may have evolved optimal habitat selection behaviors.

Fifth, studies of disturbance regimes can help guide land and resource management. For example, questions can be asked that relate to ecosystem management of forest resources. To what extent does management of forest vegetation emulate effects of various kinds of natural disturbances? Do clearcutting, partial cutting, and salvage operations have the same effects on forests as wildfires or windstorms? In one comparison, Franklin et al. (2000) reported immense differences between even-age silviculture (especially clearcutting) and natural disturbance events, such as windthrows, wildfires, and even volcanic eruptions, in terms of the types of forest components (legacies) retained or created (Table 3.4).

To meet the challenge of trying to emulate natural disturbances, some researchers have instituted a variable retention scheme in which forest openings, patches, and legacy components are kept during timber harvest operations, so as to more closely match the effects of natural disturbance events (Mitchell and Beese 2002). In the short term, the efficacy of this approach is highly dependent on the community being modified and the spatial arrangement of the retention elements, which vary greatly across taxa (Rosenvald and Lohmus 2008; Lycke et al. 2011; Bois et al. 2012; Fenton et al. 2013; Fedrowitz et al. 2014) but may increase the overall species diversity (Fedrowitz et al. 2014). Yet, as we noted above, the emulation of natural disturbances cannot now be an

Table 3.4.   Comparison of 3 types of disturbance on forest components

Legacy type	Fire	Wind	Clearcut
Snags	abundant	few	none
Logs on forest floor	common	abundant	few or none
Soil disturbance	low	patchy	high
Understory community impact	heavy	light	heavy

Source: Franklin et al. (2000)

end unto itself, because we are not actually able ac-complish this in any global sense. Managers—and even society—must also come to grips with whether resources should be expended to control natural dis-turbances. In many cases, such attempts (e.g., fire suppression, water impounds) have led to larger problems than allowing the disturbances to run their course. Therefore, we need to emulate those aspects of natural disturbances that support specific management goals and monitor the results of these actions in terms of the goals they were ex-pected to enhance.

LITERATURE CITED

Acevedo, M. F., D. L. Urban, and M. Ablan. 1995. Transition and gap models of forest dynamics. Ecological Applica-tions 5(4):1040–1055.

Alfaro, R. I., S. Taylor, R. G. Brown, and J. S. Clowater. 2001. Susceptibility of northern British Columbia forests to spruce budworm defoliation. Forest Ecology and Management 145(3):181–190.

Armstrong, G. W., W. L. Adamowicz, J. A. Beck, S. G. Cumming, and F. K. A. Schmiegelow. 2003. Coarse filter ecosystem management in a non-equilibrating forest. Forest Science 49(2):209–223.

Arroyo-Mora, J. P., G. A. Sánchez-Azofeifa, B. Rivard, J. C. Calvo, and D. H. Janzen. 2005. Dynamics in landscape structure and composition for the Chorotega region, Costa Rica, from 1960 to 2000. Agriculture, Ecosystems and Environment 106(1):27–39.

Avise, J. C. 2000. Phylogeography: The history and formation of species. Harvard University Press, Cambridge, MA.

Bahadur, K. C. 2009. Improving Landsat and IRS image classification: Evaluation of unsupervised and super-vised classification through band ratios and DEM in a mountainous landscape in Nepal. Remote Sensing 1(4):1257–1272.

Bailey, R. G. 2005. Identifying ecoregion boundaries. Environmental Management 34(1):S14–S26.

Bascompte, J., H. Possingham, and J. Roughgarden. 2002. Patchy populations in stochastic environments: Critical number of patches for persistence. American Naturalist 159:128–137.

Beissinger, S. R., J. M. Wunderle Jr., J. M. Meyers, B.-E. Sæther, and S. Engen. 2008. Anatomy of a bottleneck: Diagnosing factors limiting population growth in the Puerto Rican parrot. Ecological Monographs 78(2):185–203.

Bender, D. J., L. Tischendorf, and L. Fahrig. 2003. Using patch isolation metrics to predict animal movement in binary landscapes. Landscape Ecology 18(1):17–39.

Bergman, E. J., C. J. Bishop, D. J. Freddy, G. C. White, and P. F. Doherty Jr. 2014. Habitat management influences overwinter survival of mule deer fawns in Colorado. Journal of Wildlife Management. 78(3):448–455.

Billick, I., and K. Tonkel. 2003. The relative importance of spatial vs. temporal variability in generating a condi-tional mutualism. Ecology 84(2):289–295.

Bogaert, J., A. Farina, and R. Ceulemans. 2005. Entropy increase of fragmented habitats: A sign of human impact? Ecological Indicators 5(3):207–212.

Bois, G., L. Imbeau, and M. J. Mazerolle. 2012. Recovery time of snowshoe hare habitat after commercial thinning in boreal Quebec. Canadian Journal of Forest Research 42(1):123–133.

Bolliger, J., H. Lischke, and D. G. Green. 2005. Simulating the spatial and temporal dynamics of landscapes using generic and complex models. Ecological Complexity 2(2):107–116.

Boughton, D., and U. Malvadkar. 2002. Extinction risk in successional landscapes subject to catastrophic disturbances. Conservation Ecology 6(2):2.

Butler, K. F., and T. M. Koontz. 2005. Theory into practice: Implementing ecosystem management objectives in the USDA Forest Service. Environmental Management 35(2):138–150.

Cantwell, M. D., and R. T. T. Forman. 1993. Landscape graphs: Ecological modeling with graph theory to detect configurations common to diverse landscapes. Landscape Ecology 8(4):239–255.

Carrasco, L., X. Giam, M. Papeş, and K. S. Sheldon. 2019. Metrics of Lidar-derived 3D vegetation structure reveal contrasting effects of horizontal and vertical forest heterogeneity on bird species richness. Remote Sensing 11(7):743.

Carroll, C., B. H. McRae, and A. Brookes. 2012. Use of linkage mapping and centrality analysis across habitat gradients to conserve connectivity of gray wolf populations in western North America. Conservation Biology 26(1):78–87.

Chan-McLeod, A. C. A. 2003. Factors affecting the permeability of clearcuts to red-legged frogs. Journal of Wildlife Management 67(4):663–671.

Chen, J., J. F. Franklin, and T. A. Spies. 1992. Vegetation responses to edge environments in old-growth Douglas-fir forests. Ecological Applications 2(4):387–396.

Chen, J., J. F. Franklin, and T. A. Spies. 1993. Contrasting microclimates among clearcut, edge, and interior of old-growth Douglas-fir forest. Agricultural and Forest Meteorology 63(3–4):219–237.

Chen, J., J. F. Franklin, and T. A. Spies. 1995. Growing-season microclimatic gradients from clearcut edges into old-growth Douglas-fir forests. Ecological Applications 5(1):74–86.

Collins, B. M., D. L. Fry, J. M. Lydersen, R. Everett, and S. L. Stephens. 2017. Impacts of different land management histories on forest change. Ecological Applications 27(8):2475–2486.

Crawford, H. S., and D. T. Jennings. 1989. Predation by birds on spruce budworm Choristoneura fumiferana: Functional, numerical, and total responses. Ecology 70(1):152–163.

Crooks, K. R., A. V. Suarez, and D. T. Bolger. 2004. Avian assemblages along a gradient of urbanization in a highly fragmented landscape. Biological Conservation 115(3):451–462.

Cushman, S. A., K. McGarigal, and M. C. Neel. 2008. Parsimony in landscape metrics: Strength, universality, and consistency. Ecological Indicators 8(5):691–703.

Cushman, S. A., K. S. McKelvey, J. Hayden, and M. K. Schwartz. 2006. Gene flow in complex landscapes: Testing multiple hypotheses with causal modeling. American Naturalist 168(4):486–499.

Cushman, S. A., K. S. McKelvey, and M. K. Schwartz. 2009. Use of empirically derived source-destination models to map regional conservation corridors. Conservation Biology 23(2):368–376.

Custer, T. W., R. K. Hines, and C. M. Custer. 1996. Nest initiation and clutch size of great blue herons on the Mississippi River in relation to the 1993 flood. Condor 98(2):181–188.

D'Antonio, C. M., and P. M. Vitousek. 1992. Biological invasions by exotic grasses, the grass/fire cycle, and global change. Annual Review of Ecology and Systematics 23(1):63–87.

Degen, T., F. Devillez, and A.-L. Jacquemart. 2005. Gaps promote plant diversity in beech forests (Luzulo-Fagetum), North Vosges, France. Annals of Forest Science 62(5):429.

Delin, A. 1992. Kärlväxter i taigan i Hälsingland—deras anpassningar till kontinuitet eller störning. [Vascular plants of the taiga—adaptations to continuity or to disturbance]. Svensk Botanisk Tidskrift 86:147–176.

Dunning, J. B., Jr., D. J. Stewart, B. J. Danielson, B. R. Noon, T. L. Root, R. H. Lamberson, and E. E. Stevens. 1995. Spatially explicit population models: Current forms and future uses. Ecological Applications 5(1):3–11.

Edminster, F. C. 1938. Productivity of the ruffed grouse in New York. Transactions of the Third North American Natural Resources Conference 3:825–833.

Ehinger, L. H. 1991. Hurricane Hugo damage. Journal of Arboriculture 17(3):82–83.

Elith, J., S. J. Phillips, T. Hastie, M. Dudík, Y. E. Chee, and C. J. Yates. 2011. A statistical explanation of MaxEnt for ecologists. Diversity and Distributions 17(1):43–57.

Ellner, S. P., and G. Fussmann. 2003. Effects of successional dynamics on metapopulation persistence. Ecology 84(4):882–889.

Engstrom, R. T., S. Gilbert, M. L. Hunter Jr., D. Merriwether, G. J. Nowacki, and P. Spencer. 1999. Practical applications of disturbance ecology to natural resource management. Pp. 313–330 in R. C. Szaro, N. C. Johnson, W. T. Sexton, and A. J. Malk, eds. Ecological stewardship: A common reference for ecosystem management, vol. 2. Elsevier Science, Oxford.

Everett, R., P. Hessburg, J. Lehmkuhl, M. Jensen, and P. Bourgeron. 1994. Old forests in dynamic landscapes: Dry-site forests of eastern Oregon and Washington. Journal of Forestry 92(1):22–25.

Fahrig, L. 2017. Ecological responses to habitat fragmentation per se. Annual Review of Ecology, Evolution, and Systematics 48(1):1–23.

Fedrowitz, K., J. Koricheva, S. C. Baker, D. B. Lindenmayer, B. Palik, R. Rosenvald, W. Beese, J. F. Franklin, J. Kouki, E. Macdonald, and C. Messier. 2014. Can retention forestry help conserve biodiversity? A meta-analysis. Journal of Applied Ecology 51(6):1669–1679.

Fenton, N. J., L. Imbeau, T. Work, J. Jacobs, H. Bescond, P. Drapeau, and Y. Bergeron. 2013. Lessons learned from

12 years of ecological research on partial cuts in black spruce forests of northwestern Québec. Forestry Chronicle 89(3):350–359.

Fettig, C. J., L. A. Mortenson, B. M. Bulaon, and P. B. Foulk. 2019. Tree mortality following drought in the central and southern Sierra Nevada, US. Forest Ecology and Management 432:164–178.

Fitzsimmons, M. 2003. Effects of deforestation and reforestation on landscape spatial structure in boreal Saskatchewan, Canada. Forest Ecology and Management 174(1–3):577–592.

Foody, G. M. 2002. Status of land cover classification accuracy assessment. Remote Sensing of Environment 80(1):185–201.

Forman, R. T. T., and M. Godron. 1986. Landscape ecology. John Wiley & Sons, New York.

Forsman, E. D., R. G. Anthony, E. C. Meslow, and C. J. Zabel. 2004. Diets and foraging behavior of northern spotted owls in Oregon. Journal of Raptor Research 38(3):214–230.

Forsman, E. D., E. C. Meslow, and H. M. Wight. 1984. Distribution and biology of the spotted owl in Oregon. Wildlife Monographs 87:3–64.

Foster, D. R., D. H. Knight, and J. F. Franklin. 1998. Landscape patterns and legacies resulting from large, infrequent forest disturbances. Ecosystems 1(6):497–510.

Franklin, J. F., D. Lindenmayer, J. A. MacMahon, A. McKee, J. Magnuson, D. A. Perry, R. Waide, and D. Foster. 2000. Threads of continuity. Conservation Biology in Practice 1(1):9–16.

Frelich, L. E. 2002. Forest dynamics and disturbance regimes. Cambridge University Press, Cambridge.

Fretwell, S. D. 1972. Populations in a seasonal environment. Princeton University Press, Princeton, NJ.

Fretwell, S. D., and H. L. Lucas Jr. 1970. On territorial behavior and other factors influencing habitat distribution in birds, I: Theoretical development. Acta Biotheoretica 19(1):16–36.

Fried, E. 1975. A descriptive index of habitat shape irregularity. New York Fish and Game 22:166–167.

Fuller, R. J. 2000. Influence of treefall gaps on distributions of breeding birds within interior old-growth stands in Bialowieza Forest, Poland. Condor 102(2):267–274.

Gibson, D., T. V. Riecke, T. Keyes, C. Depkin, J. Fraser, and D. H. Catlin. 2018. Application of Bayesian robust design model to assess the impacts of a hurricane on shorebird demography. Ecosphere 9(8):e02334.

Giles, R. H., Jr., and M. K. Trani. 1999. Key elements of landscape pattern measures. Environmental Management 23(4):477–481.

Gilpin, M. E. 1990. Extinction of finite metapopulations in correlated environments. Pp. 177–186 in B. Shorrocks and I. R. Swingland, eds. Living in a patchy environment. Oxford University Press, Oxford.

Goodman, D. 1987. Consideration of stochastic demography in the design and management of biological reserves. Natural Resource Modeling 1(2):205–234.

Grant, P. R., B. R. Grant, L. F. Keller, and K. Petren. 2000. Effects of El Niño events on Darwin's finch productivity. Ecology 81(9):2442–2457.

Greenberg, C. H. 2002. Response of white-footed mice (Peromyscus leucopus) to coarse woody debris and microsite use in southern Appalachian treefall gaps. Forest Ecology and Management 164(1–3):57–66.

Grover, K. E., and M. J. Thompson. 1986. Factors influencing spring feeding site selection by elk in the Elkhorn Mountains, Montana. Journal of Wildlife Management 50(3):466–470.

Gu, W., R. Heikkila, and I. Hanski. 2002. Estimating the consequences of habitat fragmentation on extinction risk in dynamic landscapes. Landscape Ecology 17(8):699–710.

Haddad, N. M., D. R. Bowne, A. Cunningham, B. J. Danielson, D. J. Levey, S. Sargent, and T. Spira. 2003. Corridor use by diverse taxa. Ecology 84(3):609–615.

Hann, W. J., J. L. Jones, R. E. Keane, P. F. Hessburg, and R. A. Gravenmier. 1998. ICBEMP: Landscape dynamics. Journal of Forestry 96(10):10–15.

Hanski, I., and M. Gilpin. 1991. Metapopulation dynamics: Brief history and conceptual domain. Biological Journal of the Linnean Society 42(1–2):3–16.

Hargrove, W. W., F. M. Hoffman, and P. M. Schwartz. 2002. A fractal landscape realizer for generating synthetic maps. Conservation Ecology 6(1):2.

Harrison, S., B. D. Inouye, and H. D. Safford. 2003. Ecological heterogeneity in the effects of grazing and fire on grassland diversity. Conservation Biology 17(3):837–845.

Hayes, J. P., J. M. Weikel, and M. M. P. Huso. 2003. Response of birds to thinning young Douglas-fir forests. Ecological Applications 13(5):1222–1232.

He, H. S., B. E. DeZonia, and D. J. Mladenoff. 2000. An aggregation index (AI) to quantify spatial patterns of landscapes. Landscape Ecology 15(7):591–601.

Herold, A., and U. Ulmer. 2001. Stand stability in the Swiss National Forest Inventory: Assessment technique, reproducibility, and relevance. Forest Ecology and Management 145(1–2):29–42.

Herzog, F., A. Lausch, E. Müller, H. Thulke, U. Steinhardt, and S. Lehmann. 2001. Landscape metrics for assessment of landscape destruction and rehabilitation. Environmental Management 27(1):91–107.

Hlavka, C. A., and G. P. Livingston.1997. Statistical models of fragmented land cover and the effect of coarse spatial resolution on the estimation of area with satellite sensor imagery. International Journal of Remote Sensing 18(10):2253–2259.

Holt, R. D., and M. Barfield. 2003. Impacts of temporal variation on apparent competition and coexistence in open ecosystems. Oikos 101(1):49–58.

Hooper, R. G., J. C. Watson, and R. E. F. Escano. 1990. Hurricane Hugo's initial effects on red-cockaded woodpeckers in the Francis Marion National Forest. Transactions of the Fifty-Fifth North American Wildlife and Natural Resources Conference 55:220–224.

Horn, H. S. 1975. Markovian properties of forest succession. Pp. 196–211 in M. L. Cody and J. M. Diamond, eds. Ecology and evolution of communities. Harvard University Press, Cambridge, MA.

Horne, J. S., E. O. Garton, S. M. Krone, and J. S. Lewis. 2007. Analyzing animal movements using Brownian bridges. Ecology 88(9):2354–2363.

Hubbell, S. P. 2001. The unified neutral theory of biodiversity and biogeography. Princeton University Press, Princeton, NJ.

Huston, M. A. 1994. Biological diversity: The coexistence of species on changing landscapes. Cambridge University Press, Cambridge.

IIRS [Indian Institute of Remote Sensing]. 1999. BioCAP user's manual for landscape analysis and modelling biological richness. Indian Institute of Remote Sensing, National Remote Sensing Agency, Department of Space, Government of India, Dehra Dun.

Irwin, L. L., and J. M. Peek. 1983. Elk habitat use relative to forest succession in Idaho. Journal of Wildlife Management 47(3):664–672.

Jager, J. A. 2000. Landscape division, splitting index, and effective mesh size: New measures of landscape fragmentation. Landscape Ecology 15(2):115–130.

Jenerette, G. D., and J. Wu. 2000. On the definitions of scale. Bulletin of the Ecological Society of America 81(1):104–105.

Khorram, S., C. F. van der Wiele, F. H. Koch, S. A. C. Nelson, and M. D. Potts. 2016. Principles of applied remote sensing. Springer, New York.

Kie, J. G., R. T. Bowyer, M. C. Nicholson, B. B. Boroski, and E. R. Loft. 2002. Landscape heterogeneity at differing scales: Effects on spatial distribution of mule deer. Ecology 83(2):530–544.

Kinlan, B. P., and S. D. Gaines. 2003. Propagule dispersal in marine and terrestrial environments: A community perspective. Ecology 84(8):2007–2020.

Kitching, R. L., and R. A. Beaver. 1990. Patchiness and community structure. Pp. 147–176 in B. Shorrocks and I. R. Swingland, eds. Living in a patchy environment. Oxford University Press, Oxford.

Koplin, J. R. 1969. The numerical response of woodpeckers to insect prey in a subalpine forest in Colorado. Condor 71(4):436–438.

Lande, R., and G. F. Barrowclough. 1987. Effective population size, genetic variation, and their use in population management. Pp. 87–123 in M. E. Soulé, ed. Viable populations for conservation. Cambridge University Press, Cambridge.

Landguth, E. L., S. A. Cushman, M. K. Schwartz, K. S. McKelvey, M. Murphy, and G. Luikart. 2010. Quantifying the lag time to detect barriers in landscape genetics. Molecular Ecology 19(19):4179–4191.

Laurance, W. F., and G. B. Williamson. 2001. Positive feedbacks among forest fragmentation, drought, and climate change in the Amazon. Conservation Biology 15(6):1529–1535.

Lawrence, R. L., and W. J. Ripple. 2000. Fifteen years of revegetation of Mount St. Helens: A landscape-scale analysis. Ecology 81(10):2742–2752.

Lawton, R. O., and F. E. Putz. 1988. Natural disturbance and gap-phase regeneration in a wind-exposed tropical cloud forest. Ecology 69(3):764–777.

Le Comber, S. C., A. C. Spinks, N. C. Bennett, J. U. M. Jarvis, and C. G. Faulkes. 2002. Fractal dimension of African mole-rat burrows. Canadian Journal of Zoology 80(3):436–441.

Lefsky, M. A., W. B. Cohen, G. G. Parker, and D. J. Harding. 2002. Lidar remote sensing for ecosystem studies: Lidar, an emerging remote sensing technology that directly measures the three-dimensional distribution of plant canopies, can accurately estimate vegetation structural attributes, and should be of particular interest to forest, landscape, and global ecologists. BioScience 52(1):19–30.

Leopold, A. 1933. Game management. C. Scribner's Sons, New York.

Levins, R. 1969. The effects of random variation of different types on population growth. Proceedings of the National Academy of Sciences 62(4):1061–1065.

Levins, R. 1970. Extinction. Pp. 77–107 in M. Gerstenhaber, ed. Some mathematical questions in biology. American Mathematical Society, Providence, RI.

Li, H., and J. Wu. 2004. Use and misuse of landscape indices. Landscape Ecology 19(4):389–399.

Livezey, K. B. 2009. Range expansion of barred owls, part I: Chronology and distribution. American Midland Naturalist 161(1):49–56.

Luniak, M. 2004. Synurbization-adaptation of animal wildlife to urban development. Pp. 50–55 in Proceedings of the Fourth International Symposium on Urban Wildlife Conservation, University of Tucson, Tucson, AZ.

Lycke, A., L. Imbeau, and P. Drapeau. 2011. Effects of commercial thinning on site occupancy and habitat use for spruce grouse. Canadian Journal of Forest Research 41(3):501–508.

Mabry, K. E., and G. W. Barrett. 2002. Effects of corridors on home range sizes and interpatch movements of three small mammal species. Landscape Ecology 17(7):629–636.

Mandelbrot, B. B. 1975. Stochastic models for the Earth's relief, the shape and the fractal dimension of the coastlines, and the number-area rule for islands. Proceedings of the National Academy of Sciences 72(10):3825–3828.

Manly, B. F. L., L. McDonald, D. L. Thomas, T. L. McDonald, and W. P. Erickson. 2002. Resource selection by animals: Statistical design and analysis for field studies, 2nd edition. Springer Science & Business Media, New York.

Manning A. D., D. B. Lindenmayer, and H. A. Nix. 2004. Continua and Umwelt: Novel perspectives on viewing landscapes. Oikos 104(3):621–628.

Mao, X., and J. Hou. 2019. Object-based forest gaps classification using airborne LIDAR data. Journal of Forestry Research 30(2):617–627.

Marcot, B. G., and P. Z. Chinn. 1982. Use of graph theory measures for assessing diversity of wildlife habitat. Pp. 69–70 in R. Lamberson, ed. Mathematical models of renewable resources: Proceedings of the First Pacific Coast Conference on Mathematical Models of Renewable Resources. Humboldt State University, Arcata, CA.

Marcot, B. G., L. K. Croft, J. F. Lehmkuhl, R. H. Naney, C. G. Niwa, W. R. Owen, and R. E. Sandquist. 1998. Macroecology, paleoecology, and ecological integrity of terrestrial species and communities of the interior Columbia River Basin and portions of the Klamath and Great Basins. General Technical Report PNW-GTR-410. US Department of Agriculture, Forest Service, Pacific Northwest Research Station, Portland, OR.

Marcot, B. G., A. Kumar, P. S. Roy, V. P. Sawarkar, A. Gupta, and S. N. Sangma. 2002. Towards a landscape conservation strategy: Analysis of Jhum landscape and proposed corridors for managing elephants in South Garo Hills district and Nokrek area, Meghalaya. Indian Forester 128(2):207–216.

Marcot, B. G., and V. J. Meretsky. 1983. Shaping stands to enhance habitat diversity. Journal of Forestry 81(8):526–528.

Marell, A., J. P. Ball, and A. Hofgaard. 2002. Foraging and movement paths of female reindeer: Insights from fractal analysis, correlated random walks, and Levy flights. Canadian Journal of Zoology 80(5):854–865.

Masters, R. E., and K. E. M. Galley. 2007. Fire in grassland and shrubland ecosystems. Tall Timbers Fire Ecology Conference Proceedings, vol. 23. Tall Timbers Research Station, Tallahassee, FL.

McGarigal, K. 2015. FRAGSTATS help. University of Massachusetts, Amherst.

McGarigal K., and S. A. Cushman. 2005. The gradient concept of landscape structure. Pp. 112–119 in J. Wiens and M. Moss, eds. Issues and perspectives in landscape ecology. Cambridge University Press, Cambridge.

McGarigal, K., and B. J. Marks. 1995. FRAGSTATS: Spatial pattern analysis program for quantifying landscape structure. General Technical Report PNW-GTR-351. US Department of Agriculture, Forest Service, Pacific Northwest Research Station, Portland, OR.

McGarigal, K., S. Tagil, and S. A. Cushman. 2009. Surface metrics: An alternative to patch metrics for the quantification of landscape structure. Landscape Ecology 24(3):433–450.

McIntyre, N. E., and J. A. Wiens. 2000. A novel use of the lacunarity index to discern landscape function. Landscape Ecology 15(4):313–321.

McIntyre, S., and G. W. Barrett. 1992. Habitat variegation, an alternative to fragmentation. Conservation Biology 6(1):146–147.

McKelvey, K. S., and B. R. Noon. 2001. Incorporating uncertainties in animal location and map classification into habitat relationships modeling. Pp. 72–90 in C. T. Hunsaker, M. Goodchild, M. Friedl, and T. Case, eds. Spatial Uncertainty in Ecology. Springer-Verlag, New York.

McMullin, R. T., and Y. F. Wiersma. 2019. Out with old growth, in with ecological continuity: New perspectives on forest conservation. Frontiers in Ecology and the Environment 17(3):176–181.

McRae, B. H., and P. Beier. 2007. Circuit theory predicts gene flow in plant and animal populations. Proceedings of the National Academy of Sciences 104:19885–19890.

McRae, B. H., B. G. Dickson, T. H. Keitt, and V. B. Shah. 2008. Using circuit theory to model connectivity in ecology, evolution, and conservation. Ecology 89(10):2712–2724.

McRae, B. H., and V. B. Shah. 2009. Circuitscape user's guide. University of California, Santa Barbara, https://circuitscape.org.

Meentemeyer, V. 1989. Geographical perceptions of space, time, and scale. Landscape Ecology 3(3–4):163–173.

Metlen, K. L., C. N. Skinner, D. R. Olson, C. Nichols, and D. Borgias. 2018. Regional and local controls on historical fire regimes of dry forests and woodlands in the Rogue

River Basin, Oregon. Forest Ecology and Management 430:43–58.

Millar, C. I., N. L. Stephenson, and S. L. Stephens. 2007. Climate change and forests of the future: Managing in the face of uncertainty. Ecological Applications 17(8):2145–2151.

Miller, M. W., and T. D. Nudds. 1996. Prairie landscape change and flooding in the Mississippi River Valley. Conservation Biology 10(3):847–853.

Mitchell, S. J., and W. J. Beese. 2002. The retention system: Reconciling variable retention with the principles of silvicultural systems. Forestry Chronicle 78(3):397–403.

Moeur, M. 1997. Spatial models of competition and gap dynamics in old-growth *Tsuga heterophylla / Thuja plicata* forests. Forest Ecology and Management 94(1–3):175–186.

Morrison, M. L. 2009. Restoring wildlife: Ecological concepts and practical applications. Island Press, Washington, DC.

Morrison, M.L., B. G. Marcot, and R. W. Mannan. 2006. Wildlife-habitat relationships. Island Press, Washington, DC.

Mortberg, U. M. 2001. Resident bird species in urban forest remnants: Landscape and habitat perspectives. Landscape Ecology 16(3):193–203.

Moser, D., H. G. Zechmeister, C. Plutzar, N. Sauberer, T. Wrbka, and G. Grabherr. 2002. Landscape patch shape complexity as an effective measure for plant species richness in rural landscapes. Landscape Ecology 17(7):657–669.

Nams, V. O., and M. Bourgeois. 2004. Fractal analysis measures habitat use at different spatial scales: An example with American marten. Canadian Journal of Zoology 82(11):1738–1747.

Neel, M. C., K. McGarigal, and S. A. Cushman. 2004. Behavior of class-level landscape metrics across gradients of class aggregation and area. Landscape Ecology 19(4):435–455.

Ohman, K., and T. Lamas. 2005. Reducing forest fragmentation in long-term forest planning by using the shape index. Forest Ecology and Management 212(1–3):346–357.

Opdam, P., J. Verboom, and R. Pouwels. 2003. Landscape cohesion: An index for the conservation potential of landscapes for biodiversity. Landscape Ecology 18(2):113–126.

Orrock, J. L., B. J. Danielson, M. J. Burns, and D. J. Levey. 2003. Spatial ecology of predator-prey interactions: Corridors and patch shape influence seed predation. Ecology 84(10):2589–2599.

Parresol, B. R., and J. McCollum. 1997. Characterizing and comparing landscape diversity using GIS and a contagion index. Journal of Sustainable Forestry 5(1–2):249–261.

Parsons, W. F., D. H. Knight, and S. L. Miller. 1994. Root gap dynamics in lodgepole pine forest: Nitrogen transformations in gaps of different size. Ecological Applications 4(2):354–362.

Patriquin, K. J., and R. M. R. Barclay. 2003. Foraging by bats in cleared, thinned, and unharvested boreal forest. Journal of Applied Ecology 40(4):646–657.

Patterson, T. A., L. Thomas, C. Wilcox, O. Ovaskainen, and J. Matthiopoulos. 2008. State-space models of individual animal movement. Trends in Ecology & Evolution 23:87–94.

Patton, D. R. 1975. A diversity index for quantifying habitat edge. Wildlife Society Bulletin 3(4):171–173.

Peterson, D. L., and V. T. Parker. 1998. Dimensions of scale in ecology, resource management, and society. Pp. 499–522 in D. L. Peterson and V. T. Parker, eds. Ecological scale: Theory and applications. Columbia University Press, New York.

Phillips, S. J., and M. Dudík. 2008. Modeling of species distributions with MaxEnt: New extensions and a comprehensive evaluation. Ecography 31(2):161–175.

Pielou, E. C. 2008. After the ice age: The return of life to glaciated North America. University of Chicago Press, Chicago.

Pierson, E. D., T. Elmqvist, W. E. Rainey, and P. A. Cox. 1996. Effects of tropical cyclonic storms on flying fox populations on the South Pacific islands of Samoa. Conservation Biology 10(2):438–451.

Plumb, R. E., K. M. Gordon, and S. H. Anderson. 2003. Pronghorn use of a wildlife underpass. Wildlife Society Bulletin 31(4):1244–1245.

Pullinger, M. G., and C. J. Johnson. 2010. Maintaining or restoring connectivity of modified landscapes: Evaluating the least-cost path model with multiple sources of ecological information. Landscape Ecology 25(10):1547–1560.

Quinn, J. F., and A. Hastings. 1987. Extinction in subdivided habitats. Conservation Biology 1(3):198–208.

Rojo, J. M. and S. S. Orois. 2005. A decision support system for optimizing the conversion of rotation forest stands to continuous cover forest stands. Forest Ecology and Management 207(1–2):109–120.

Rosenberg, D. K., and R. G. Anthony. 1992. Characteristics of northern flying squirrel populations in young second- and old-growth forests in western Oregon. Canadian Journal of Zoology 70(1):161–166.

Rosenberg, D. K., B. R. Noon, and E. C. Meslow. 1997. Biological corridors: Form, function, and efficacy. BioScience 47(10):677–687.

Rosenvald, R., and A. Lohmus. 2008. For what, when, and where is green-tree retention better than clear-cutting? A review of the biodiversity aspects. Forest Ecology and Management 255(1):1–15.

Roy, P. S., and S. Tomar. 2000. Biodiversity characterization at the landscape level using geospatial modelling technique. Biological Conservation 95(1): 95–109.

Sadowski, F. G., J. A. Sturdevant, and R. A. Rowntree. 1987. Testing the consistency for mapping urban vegetation with high-altitude aerial photographs and landsat MSS data. Remote Sensing of Environment 21(2):129–141.

Sakai, H. F., and B. R. Noon. 1993. Dusky-footed woodrat abundance in different-aged forests in northwestern California. Journal of Wildlife Management 57(2): 373–382.

Salajanu, D., and C. E. Olson. 2001. The significance of spatial resolution: Identifying forest cover from satellite data. Journal of Forestry 99(6):32–38.

Sands, J. P., M. J. Schnupp, T. W. Tienert, S. J. DeMaso, F. Hernandez, L. A. Brennan, D. Rollins, and R. M. Perez. 2013. Tests of an additive harvest mortality model for northern bobwhite (Colinus virginianus) harvest management in Texas, USA. Wildlife Biology 19(1):12–18.

Sawyer, H., R. M. Nielson, F. G. Lindzey, L. Keith, J. H. Powell, and A. A. Abraham. 2007. Habitat selection of Rocky Mountain elk in a nonforested environment. Journal of Wildlife Management 71(3):868–874.

Schmidt, K. A., J. M. Earnhardt, J. S. Brown, and R. D. Holt. 2000. Habitat selection under temporal heterogeneity: Exorcizing the Ghost of Competition Past. Ecology 81(9):2622–2630.

Schnitzer, S. A., and W. P. Carson. 2001. Treefall gaps and the maintenance of species diversity in a tropical forest. Ecology 82(4):913–919.

Schnupp, M. J., and D. S. DeLaney. 2012. Funding research as an investment for improving management. Pp. 99–118 in J. P Sands, S. J. DeMaso, M. J. Schnupp, and L. A. Brennan, eds. Wildlife science: Connecting research with management. CRC Press, Taylor & Francis Group, Boca Raton, FL.

Schooley, R. L., and J. A. Wiens. 2003. Finding habitat patches and directional connectivity. Oikos 102(3):559–570.

Schreiber, R. W., and E. A. Schreiber. 1984. Central Pacific seabirds and the El Niño southern oscillation: 1982 to 1983 perspectives. Science 225(4663):713–716.

Selva, S. B. 1994. Lichen diversity and stand continuity in the northern hardwoods and spruce-fir of northern New England and western New Brunswick. Bryologist 97(4):424–429.

Selva, S. B. 2003. Using calicioid lichens and fungi to assess ecological continuity in the Acadian Forest ecoregion of the Canadian Maritimes. Forestry Chronicle 79(3):550–558.

Silva, M., L. A. Harling, S. A. Field, and K. Teather. 2003. The effects of habitat fragmentation on amphibian species richness of Prince Edward Island. Canadian Journal of Zoology 81(4):563–573.

Simard, A. J., D. A. Haines, and W. A. Main. 1985. Relations between El Niño / Southern Oscillation anomalies and wildland fire activity in the United States. Agricultural and Forest Meteorology 36(2):93–104.

Simberloff, D., J. A. Farr, J. Cox, and D. W. Mehlman. 1992. Movement corridors: Conservation bargains or poor investments? Conservation Biology 6(4):493–504.

Šimová, P., and K. Gdulová. Landscape indices behavior: A review of scale effects. Applied Geography 34:385–394.

Soderstrom, B., and T. Part. 2000. Influence of landscape scale on farmland birds breeding in semi-natural pastures. Conservation Biology 14(2):522–533.

Solé, R. V., D. Alonso, and J. Saldaña. 2004. Habitat fragmentation and biodiversity collapse in neutral communities. Ecological Complexity 1(1):65–75.

Spear, S. F., N. Balkenhol, M. J. Fortin, B. H. McRae, and K. Scribner. 2010. Use of resistance surfaces for landscape genetic studies: Considerations for parameterization and analysis. Molecular Ecology 19(17):3576–3591.

Spiller, D. A., and A. A. Agrawal. 2003. Intense disturbance enhances plant susceptibility to herbivory: Natural and experimental evidence. Ecology 84(4):890–897.

Stamps, J. A., M. Buechner, and V. V. Krishnan. 1987. The effects of edge permeability and habitat geometry on emigration from patches of habitat. American Naturalist 129(4):533–552.

Stapp, P., M. F. Antolin, and M. Ball. 2004. Patterns of extinction in prairie dog metapopulations: Plague outbreaks follow El Niño events. Frontiers in Ecology and the Environment 2(5):235–240.

Stauffer, D., and A. Aharony. 2018. Introduction to percolation theory. Taylor & Francis, London.

St. Clair, C. C. 2003. Comparative permeability of roads, rivers, and meadows to songbirds in Banff National Park. Conservation Biology 17(4):1151–1160.

Stoms, D. 1994. Scale dependence of species richness maps. Professional Geographer 46(3):346–358.

Sullivan, T. P., D. S. Sullivan, P. M. Lindgren, and J. O. Boateng. 2002. Influence of conventional and chemical thinning on stand structure and diversity of plant and mammal communities in young lodgepole pine forest. Forest Ecology and Management 170(1–3):173–187.

Summerville, K. S., and T. O. Crist. 2003. Determinants of lepidopteran community composition and species diversity in eastern deciduous forests: Roles of season, eco-region, and patch size. Oikos 100(1):134–148.

Suzuki, N., and J. P. Hayes. 2003. Effects of thinning on small mammals in Oregon coastal forests. Journal of Wildlife Management 67(2):352–371.

Sverdrup-Thygeson, A., and D. B. Lindenmayer. 2003. Ecological continuity and assumed indicator fungi in boreal forest: The importance of the landscape matrix. Forest Ecology and Management 174(1–3):353–363.

Taillie, P. J., and C. E. Moorman. 2019. Marsh bird occupancy along the shoreline-to-forest gradient as marshes migrate from rising sea level. Ecosphere 10(1):e02555.

Tallmon, D. A., E. S. Jules, N. J. Radke, and L. S. Mills. 2003. Of mice and men and *Trillium*: Cascading effects of forest fragmentation. Ecological Applications 13(5):1193–1203.

Tempel, D. J., J. J. Keane, R. J. Gutiérrez, J. D. Wolfe, G. M. Jones, A. Koltunov, C. M. Ramirez, W. J. Berigan, C. V. Gallagher, T. E. Munton, and P. A. Shaklee. 2016. Meta-analysis of California spotted owl (*Strix occidentalis occidentalis*) territory occupancy in the Sierra Nevada: Habitat associations and their implications for forest management. Condor: Ornithological Applications 118(4):747–765.

Thomas, J. W., H. J. Black, R. J. Scherzinger, and R. J. Pedersen. 1979. Deer and elk. Pp. 104–127 *in* J. W. Thomas, technical ed. Wildlife habitats in managed forests: The Blue Mountains of Oregon and Washington. Agricultural Handbook 533. US Department of Agriculture, Washington, DC.

Thomas, J. W., E. D. Forsman, J. B. Lint, E. C. Meslow, B. R. Noon, and J. Verner. 1990. A conservation strategy for the northern spotted owl: Report of the Interagency Scientific Committee to address the conservation of the northern spotted owl. Unpublished interagency document.

Thomas, J. W., D. A. Leckenby, M. Henjum, R. J. Pedersen, and L. D. Bryant. 1988. Habitat-effectiveness index for elk on Blue Mountain winter ranges. General Technical Report PNW-GTR-218. US Department of Agriculture, Forest Service, Pacific Northwest Research Station, Portland, OR.

Thompson, J. N. 1999. Specific hypotheses on the geographic mosaic of coevolution. American Naturalist 153:S1–S14.

Tibell, L. 1992. Crustose lichens as indicators of forest continuity in boreal coniferous forests. Nordic Journal of Botany 12(4):427–450.

Tischendorf, L., D. J. Bender, and L. Fahrig. 2003. Evaluation of patch isolation metrics in mosaic landscapes for specialist vs. generalist dispersers. Landscape Ecology 18(1):41–50.

Turner, M. G., W. H. Romme, and D. B. Tucker. 2003. Surprises and lessons from the 1988 Yellowstone fires. Frontiers in Ecology and the Environment 1(7):351–358.

Upadhyay, S., A. Roy, M. Ramprakash, J. Idiculla, A. S. Kumar, and S. Bhattacharya. 2017. A network theoretic study of ecological connectivity in western Himalayas. Ecological Modelling 359:246–257.

Urban, D., and T. Keitt. 2001. Landscape connectivity: A graph-theoretic perspective. Ecology 82(5):1205–1218.

Valentine, K. W. G. 1981. How soil map units and delineations change with survey intensity and map scale. Canadian Journal of Soil Science 61(4):535–551.

Vergara, P. M. 2011. Matrix-dependent corridor effectiveness and the abundance of forest birds in fragmented landscapes. Landscape Ecology 26(8):1085.

Vergara, P. M., C. G. Pérez-Hernández, I. J. Hahn, and J. E. Jiménez. 2013. Matrix composition and corridor function for austral thrushes in a fragmented temperate forest. Landscape Ecology 28(1):121–133.

Vilella, F. J., and E. R. Garcia. 1995. Post-hurricane management of the Puerto Rican parrot. Pp. 618–621 *in* J. A. Bissonette and P. R. Krausman, eds. Integrating people and wildlife for a sustainable future: Proceedings of the First International Wildlife Management Congress. The Wildlife Society, Bethesda, MD.

Wade, A. A., K. S. McKelvey, and M. K. Schwartz. 2015. Resistance-surface-based wildlife conservation connectivity modeling: Summary of efforts in the United States and guide for practitioners. General Technical Report RMRS-GTR-333. US Department of Agriculture, Forest Service, Rocky Mountain Research Station, Fort Collins, CO.

Wagner, H. H., and M. J. Fortin. 2005. Spatial analysis of landscapes: Concepts and statistics. Ecology 86(8):1975–1987.

Walker, L. R. 1991. Tree damage and recovery from Hurricane Hugo in Luquillo experimental forest, Puerto Rico. Biotropica 23(4):379–385.

Walker, R. S., A. J. Novaro, and L. C. Branch. 2003. Effects of patch attributes, barriers, and distance between patches on the distribution of a rock-dwelling rodent (*Lagidium viscacia*). Landscape Ecology 18(2):185–192.

Walters, J. R. 1991. Application of ecological principles to the management of endangered species: The case of the red-cockaded woodpecker. Annual Review of Ecology and Systematics 22(1):505–523.

Webb, S. L. 1989. Contrasting windstorm consequences in two forests, Itasca State Park, Minnesota. Ecology 70(4):1167–1180.

Weinkle, J., R. Maue, and R. Pielke Jr. 2012. Historical global tropical cyclone landfalls. Journal of Climate 25(13):4729–4735.

Weltzin, J. F., M. E. Loik, S. Schwinning, D. G. Williams, P. A. Fay, B. M. Haddad, J. Harte, T. E. Huxman, A. K. Knapp, and G. Lin. 2003. Assessing the response of terrestrial ecosystems to potential changes in precipitation. BioScience 53(10):941–952.

Wiemers, D. W., T. E. Fulbright, D. B. Wester, J. A. Ortega-S, G. A. Rasmussen, D. G. Hewitt, and M. W. Hellickson. 2014. Role of thermal environment in habitat selection by male white-tailed deer during summer in Texas. Wildlife Biology 20(1):47–57.

Wiens, J. A. 1976. Population response to patchy environments. Annual Review of Ecology and Systematics 7(1):81–120.

Wiens, J. A. 1989. Spatial scaling in ecology. Functional Ecology 3(4):385–397.

Wiens, J. A. 2016. Ecological challenges and conservation conundrums: Essays and reflections for a changing world. John Wiley & Sons, West Sussex, UK.

Williams, D. W., and A. W. Liebhold. 2000. Spatial synchrony of spruce budworm outbreaks in eastern North America. Ecology 81(10):2753–2766.

Withers, M. A., and V. Meentemeyer. 1999. Concepts of scale in landscape ecology. Pp. 205–252 in J. M. Klopatek and R. H. Gardner, eds. Landscape ecological analysis: Issues and applications. Springer-Verlag, New York.

Wu, J., and O. L. Loucks. 1995. From balance of nature to hierarchical patch dynamics: A paradigm shift in ecology. Quarterly Review of Biology 70(4):439–466.

Wunderle, J. M., Jr. 1995. Responses of bird populations in a Puerto Rican forest to Hurricane Hugo: The first 18 months. Condor 97(4):879–896.

Zabel, C. J., K. McKelvey, and J. P. Ward Jr. 1995. Influence of primary prey on home-range size and habitat use patterns of northern spotted owls (Strix occidentalis caurina). Canadian Journal of Zoology 73(3):433–439.

Zonneveld, I. S. 1979. Land evaluation and land(scape) science, 2nd edition. ITC textbook of photo-interpretation, vol. 7. International Institute for Aerial Survey and Earth Sciences, Enchede, Netherlands.

4 — The Evolutionary Perspective
Linking Habitat to Population

Nothing in biology makes sense except in the light of evolution.

Dobzhansky (1973:125)

Animal ecology is more than natural history. That is, it goes beyond simply documenting the occurrence and distribution of organisms and their environments, although that is an essential building block for describing, understanding, and predicting the dynamics of biotic systems. More fully, the study of animal ecology is challenged by interpreting recent or current conditions in the context of complex and vast histories that led to those conditions. Ultimately, this is an evolutionary perspective.

As we laid out in Chapter 1, some aspects of the conservation of animals pertain to knowing and providing for their evolutionary legacy through a perspective of how animals both react to and become adapted to a set of environmental conditions. This, then, helps describe and explain adaptive advantages (or disadvantages) and conditions that allow their persistence, including the environments they use and select which, in turn, define their habitats.

In this chapter, we explore the underlying concepts of an evolutionary perspective, to be followed later (in Chapter 6) by their implications for practical management guidelines. Here, we review terms and concepts pertaining to such an ecosystem and its evolutionary perspective.

Individuals, Populations, and Evolution

As we noted in Chapter 1, populations are essentially the unit of evolution and adaptive change. It is at the population level where individual variation is expressed—that is, the relative frequencies of phenotypes and behaviors among individuals that add to, or take from, the advantages of reproduction and survival in a given environmental condition under changing conditions. In a sense, individual variation is the fuel for evolution and adaptation. If all members of a population were identical, there would be no advantage given to any, other than through happenstance and random genetic mutation. Yet the genetics of reproduction essentially ensures that some variations will occur, at least through random mutation of alleles, if nothing else.

In instances of isolated, purely inbred populations, such variation may be minimal. In these cases, changes at the population level might occur only through random genetic drift (Allendorf 1986), and populations may be unable to sustain themselves or adapt to current or changing environments. This has happened to some small inbred populations of desert bighorn sheep (*Ovis canadensis nelsoni*) (Hedrick 2014). Small-population dynamics (Wootton and Pfister 2013) that lead to local extinctions—

called an "extinction risk vortex"—can include demographic stochasticity (random variation in rates of survival and reproduction), Allee effects (threshold population sizes fostering successful reproduction), and environmental stochasticity (random environmental fluctuations). But in most healthy animal populations, individual variations in phenotypes and behavior help ensure that the engine of evolution and adaptation can proceed.

We can define adaptation as a change in the modal frequency of a morphology or behavior within a population that maintains or confers a relatively greater degree of survival or fecundity on some individuals over other individuals. But what is the speed of evolution and adaptation? At varying scales, these can proceed at vastly different rates. Changes in entire species lineages typically occur slowly, through shifts in the average morphology, anatomy, physiology, and behavior of individuals. This typically occurs over eons, millennia, or perhaps just centuries, depending on the organism's generational time frame and modifications brought about by changes in environmental conditions.

Adaptation within a lineage, however, can occur far more rapidly, such as with altered mating signals among individuals, as has occurred in white-crowned sparrows (*Zonotrichia leucophrys*). A study by Derryberry (2009) in the western United States suggested that, over a span of 35 years, variations in the density of vegetation affected the acoustic properties of the environment, rendering an advantage to some white-crowned sparrows with different bill sizes and vocalization abilities, which came to dominate the modal phenotype of the population. Adaptation within this lineage occurred relatively quickly.

Most biologists view evolution as an immensely slow process of changes in the morphology, anatomy, or physiology of organisms, which occur on the order of millennia and eons. This may be true for some taxonomic lineages—and higher taxonomic levels, such as families and orders, tend to change more slowly than lower levels, such as genera and species. There are startling new cases, however, of quick mi-croevolutionary shifts caused by rapid alterations in environmental conditions due to human activities (e.g., Pergams and Ashley 1999; Pergams et al. 2003; Sayama et al. 2003). These cases clearly demonstrate that the adaptation of forms and modal genotypes to changing conditions can occur over centuries, decades, and, in some cases, during the course of just a few years, as demonstrated by Losos et al. (1997) and Harmon et al. (2010). This puts an entirely new perspective on what managers ought to be focusing on when providing habitats for wildlife and projecting the effects of management activities in an evolutionary context.

What Is an Evolutionary Perspective?

The discussion thus far has put some contours on the question, "What, exactly, is an evolutionary perspective?" In general, it explains how habitat and resource use, as well as selection patterns, have developed. It also reveals the relatedness of, or differences among, populations of importance to genetic and phenotypic diversity of a species, including subspecies, evolutionarily significant units, distinct population segments, significant portions of a species' range, and identifiable groups of other individuals (see Chapter 6). An evolutionary perspective also pertains to understanding long-term conditions and alterations in a species' lineage, as well as short-term diversity that contributes to the degree to which populations and segments of a species may adapt to new or changing conditions.

Evolution is a unifying theory in biology and the life sciences. When coupled with biogeography, it provides tremendous insight into why populations of wild animals and native plants are found where they are today (Lomolino et al. 2004). Much of contemporary habitat and population management, however, focuses on the more immediate concerns of mitigation, conservation, and restoration. Nonetheless, environments will continue to change, climates will vary stochastically, and disturbance events, such as major fires, will recur. Nature is not

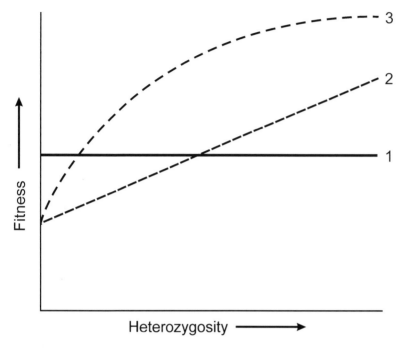

Figure 4.1   A theoretical relationship between the fitness of an individual animal and genetic heterozygosity, illustrating directional selection (*1*), neutral selection (*2*), and genetic drift (*3*). (Reproduced from Frankel and Soule (1981:48), with the permission of Cambridge University Press)

static. Thus, a particular philosophy of management holds that we should view wildlife in an evolutionary context, an idea proposed nearly 40 years ago by Frankel and Soulé (1981) that has been overlooked or sidestepped far too often. Although specific outcomes of future evolution are unpredictable (genetic engineering aside), a key focus of management and conservation should be to provide environmental conditions that permit natural evolutionary processes to proceed, including species interactions and the role of abiotic factors in selection pressures. The ebb and flow of taxonomic lineages—including the emergence of subspecies, species, and higher taxonomic groupings—is afforded only by providing a diverse and stochastic array of environmental conditions and gene pools of organisms through space and over time.

Adaptations and genetic changes in organism lineages can be caused by many factors and can take many forms. A few examples are directional selection, in which the modal genotype and phenotype of a population shifts consistently in 1 direction, such as to longer bill length to better exploit some

new resource; neutral selection, in which no single phenotype is favored but modes or variations in genomes can still change; character divergence, such as between different morphs, that results in 2 modal phenotypes; genetic drift, which may produce new modal phenotypes from random reselection in small isolated populations; and inbreeding depression in small isolated populations, which shifts modal genotypes and lowers the levels of mean fitness (Fig. 4.1).

Setting the Stage

Some situations that could contribute to adaptive changes and evolutionary responses in a population include

- directional selection—a monotonic shift in a more advantageous, modal phenotype and genotype of a population; and
- character divergence—where typically 2, atypically more than 2, phenotypes develop from 1 parent population stock, usually because of differing environmental conditions.

Probably not contributing to adaptive changes, but still of key consideration, are conditions of

- neutral selection—where some morphological or behavioral characteristic is neither advantageous nor disadvantageous;
- genetic drift—usually monotonic changes in the modal genome of a small, isolated, and sometimes inbred population; and
- inbreeding depression—an occurrence in small isolated populations where breeding pairs are closely related.

Further, behavioral flexibility is a concept that is often conflated with adaptation. Flexibility is the capability of individual organisms to modify their behaviors (including reproductive behavior) in the face of new environmental conditions (Beever et al. 2017). It can be indexed by the degree of behavioral plasticity of individual organisms and measured as the degree of organisms' resilience to environmental stressors, as well as the level at which they maintain reproductive and survival fitness in the face of acute disturbances and chronic shifts in environmental conditions. Note that flexibility is not "adaptation" per se, as this term is sometimes erroneously used, but, in some circumstances, behavioral flexibility might set the stage for adaption.

Why Is an Evolutionary Perspective Pertinent?

There are several reasons why an evolutionary perspective is pertinent to species conservation, recovery, and management.

- The patterns and dynamics of resource and habitat selection by organisms are expressions of evolutionary adaptations. Understanding them is key to effective management and conservation, such as ensuring that an introduction or translocation is made of a genetic stock appropriate to a particular environmental condition.

- Genetic relationships at various taxonomic levels—particularly genus, superspecies, species, subspecies, and population—and how they developed convey information on which entities to select, study, and understand for a potential management, conservation, or restoration focus. You cannot conserve biodiversity if you do not know the pieces.
- An understanding of the conditions that permit ongoing or potential evolutionary processes may be important for species-specific habitat management and conservation.

Managing for Evolutionary Time Frames

Embarking on management strategies that consider evolutionary time frames can be daunting. We are not suggesting that evolutionary changes could be predicted—this is not yet possible given our current ecological knowledge (Bennett 1996). Rather, we are advocating that the base conditions from which evolution proceeds could be provided via management actions and conservation policies. Ashley et al. (2003:115) described such an approach as "evolutionarily enlightened management." In this way, the array of ecological domains—the spectra of conditions and variations thereof—can be provided over time, which forms the foundation for potential long-term evolutionary change in species and their populations (Lankau et al. 2011).

Rice and Emery (2003:469) cautioned, "While there may not be any universal rules regarding the adaptive potential of species, an understanding of the various processes involved in microevolution will increase the short- and long-term success of conservation and restoration efforts." Microevolutionary processes can indeed have important consequences for species conservation, management, and restoration (Ashley et al. 2003). Thus evolutionary time frames span both long- and short-term conditions, including a microevolutionary potential in the face of rapid environmental change. Keeping these factors in mind is especially useful for considering the possible impacts of climate change on populations of wild animals (Folke et al. 2004; Mawdsley et al. 2009; Bellard et al. 2012). In Chapter 6, we further discuss specific conditions lending to provisions for the evolutionary potential of populations and species.

What Is an Ecosystem Context?

An evolutionary perspective necessitates understanding organisms in the circumstances and conditions of their ecosystems. An ecosystem consists of organisms of various taxonomic designations and levels of biological organization, along with their interactions with each other and among abiotic conditions and processes. An ecosystem is more than a mere collection of populations (organisms of the same species in some predetermined or predefined area), species assemblages (groups of species of a particular taxa), or communities (various species and their interactions). Understanding wildlife in an ecosystem context also requires understanding (1) population dynamics, including demographic and genetic variations; (2) the evolutionary context of organisms, populations, and species, including the contribution of genetic variation to persistence of species lineages, mechanisms of speciation and hybridization, and selection for adaptive traits; (3) interactions among species that affect their persistence and can potentially influence community structure, including obligate mutualisms, such as pollination or dispersal vectors, predation, competition, and others; and (4) the influence of an abiotic environment on the vitality of organisms (organism health and realized fitness) and populations (viability), including how disturbance mechanisms operate and how organisms respond. Habitat ecology plays a key role in many of these facets of an ecosystem context, but it needs to be subsumed into a broader ecosystem context that links it with population ecology (see Chapter 1 and elsewhere throughout this book).

An ecosystem context also requires understanding the role of humans in modifying environments, habitats, and populations of wild animals. The question of whether humans are a part of the ecosystem is not particularly useful. Rather, there is an abundance of evidence clearly showing that *Homo sapiens* can impact ecosystems. After all, humans have created an entirely new (and unofficial) geologic epoch, the Anthropocene, defined by the degree to which we are altering major biogeochemical cycles and conditions on this planet (Steffen et al. 2011).

Nevertheless, the admission that humans are part of the ecosystem should not be used as a rationale to justify the philosophy that abuse of resources is acceptable, because anything that humans do is "natural." A more useful and challenging question is, "To what extent do we want our actions to modify environments, habitats, and wildlife populations?" Also, "To what extent do we need to change our own resource consumption and land use habits to meet our own goals for conservation?" And, perhaps most difficult of all, "How many humans should occupy a specific area?" Answers to such questions are critical, both for successful wildlife management *and* for providing people with sustainable resources. These are 2 inextricable sides of the same coin. An ecosystem context for studying and managing wildlife should prompt such questions (Fig. 4.2).

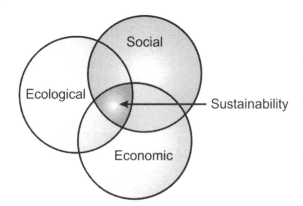

Figure 4.2 Venn diagram illustrating 3 major components of ecosystem management in the context of sustainability. (Reproduced from Haufler et al. (2002), with the permission of The Wildlife Society)

Our major influence on Earth's environments also leads to the formation of novel ecosystems, which are new combinations of species induced by rapid changes in their distributions, which, in turn, are caused by swift changes in environmental conditions and disturbances (Hobbs et al. 2009; Collier 2015). Clement and Standish (2018) addressed the technical, social, and ethical issues associated with the emergence of novel ecosystems in the Anthropocene, as well as how governance mechanisms could address such issues. They raised many of the same questions posed above. What *should* our influence be on natural systems? How should we be affecting ecosystems and, for that matter, the evolutionary potential of populations and entire species lineages?

A related question that also goes to the heart of management policy and environmental ethics is, "What is natural?" In areas only relatively recently occupied and changed by humans' technological presence and high human density, such as much of western North America, the question may seem silly. It is obvious that ancient forests and ungrazed grasslands are natural, and tree farms and croplands are not. A significant portion of the world lives in ecosystems that have been affected for millennia by human activities, however, often so much so that the question itself has little relevance.

Conclusions

It is a tricky business to predict adaptations and evolutionary changes, beyond comparing current environmental conditions that are used and selected by individuals, in order to project future conditions. Although we have worked out some of the main genetic mechanisms of evolution, our predictive power in this field remains weak, at best. We can describe and explain post hoc observed patterns of adaptation in terms of the outcome of differential reproductive and survival fitness among individuals in a population, but predicting what genetic and phenotypic changes come next remains mostly elusive.

LITERATURE CITED

Allendorf, F. W. 1986. Genetic drift and the loss of alleles versus heterozygosity. Zoo Biology 5:181–190.

Ashley, M. V., M. F. Willson, O. R. W. Pergams, D. J. O'Dowd, S. M. Gende, and J. S. Brown. 2003. Evolutionary enlightened management. Biological Conservation 111:115–123.

Beever, E. A., L. E. Hall, J. M. Varner, A. E. Loosen, J. B. Dunham, M. K. Gahl, F. A. Smith, and J. J. Lawler. 2017. Behavioral flexibility as a mechanism for coping with climate change. Frontiers in Ecology and Evolution 15(6):299–308, doi:10.1002/fee.1502.

Bellard, C., C. Bertelsmeier, P. Leadly, W. Thuiller, and F. Courchamp. 2012. Impacts of climate change on the future of biodiversity. Ecology Letters 15:365–377.

Bennett, K. D. 1996. Evolution and ecology: The pace of life. Cambridge University Press, New York.

Clement, S., and R. J. Standish. 2018. Novel ecosystems: Governance and conservation in the age of the Anthropocene. Journal of Environmental Management 208:36–45.

Collier, M. J. 2015. Novel ecosystems and social-ecological resilience. Landscape Ecology 30(8):1363–1369.

Derryberry, E. P. 2009. Ecology shapes birdsong evolution: Variation in morphology and habitat explains variation in white-crowned sparrow song. American Naturalist 174(1):24–33.

Dobzhansky, T. 1973. Nothing in biology makes sense except in light of evolution. American Biology Teacher 35:125–129.

Folke, C., S. Carpenter, B. Walker, M. Scheffer, T. Elmqvist, L. Gunderson, and C. S. Holling. 2004. Regime shifts, resilience, and biodiversity in ecosystem management. Annual Reviews in Ecology and Systematics 35:557–581.

Frankel, O. H., and M. E. Soulé. 1981. Conservation and evolution. Cambridge University Press, Cambridge.

Harmon, L. J., J. B. Losos, T. J. Davies, R. G. Gillespie, J. L. Gittleman, W. B. Jennings, K. H. Kolzak, M. A. McPeek, F. Moreno-Roark, T. J. Near, A. Purvis, R. E. Ricklefs, D. Schluter, J. A. Schulte II, O. Sheehausen, B. L. Sidlauskas, O. Torres-Carvajal, J. T. Weir, and A. O. Mooers. 2010. Early bursts of body size and shape are rare in comparative data. Evolution 64:2385–2396.

Haufler, J. B., R. K. Baydack, H. Campa III, B. J. Kernohan, C. Miller, L. J. O'Neil, and L. Waits. 2002. Performance measures for ecosystem management and ecological sustainability. Technical Review 02-1. The Wildlife Society, Bethesda, MD.

Hedrick, P. W. 2014. Conservation genetics and the persistence and translocation of small populations: Bighorn sheep populations as examples. Animal Conservation 17(2):106–114.

Hobbs, R. J., E. Higgs, and J. A. Harris. 2009. Novel ecosystems: Implications for conservation and restoration. Trends in Ecology & Evolution 24(11):599–605.

Lankau, R., P. S. Jorgensen, D. J. Harris, and A. Sih. 2011. Incorporating evolutionary principles into environmental management and policy. Evolutionary Applications 4:315–325.

Lomolino, M. V., D. F. Sax, and J. H. Brown, eds. 2004. Foundations of biogeography: Classic papers and commentaries. University of Chicago Press, Chicago.

Losos, J. B., K. T. Warheit, and T. W. Schoener. 1997. Adaptive differentiation following experimental island colonization in Anolis lizards. Nature 387:70–73.

Mawdsley, J. R., R. O'Malley, and D. S. Ojima. 2009. A review of climate-change adaptation strategies for wildlife diversity and conservation. Conservation Biology 23:1080–1089.

Pergams, O. R. W., and M. V. Ashley. 1999. Rapid morphological change in Channel Island deer mice. Evolution 53:1573–1581.

Pergams, O. R. W., W. M. Barnes, and D. Nyberg. 2003. Rapid change of mouse mitochondrial DNA. Nature 423:397.

Rice, K. J., and N. C. Emery. 2003. Managing microevolution: Restoration in the face of global change. Frontiers in Ecology and the Environment 1:469–478.

Sayama, H., L. Kaufman, and Y. Bar-Yam. 2003. Spontaneous pattern formation and genetic diversity in habitats with irregular geographical features. Conservation Biology 17:893–900.

Steffen, W., Å. Persson, L. Deutsch, J. Zalasiewicz, M. Williams, K. Richardson, C. Crumley, P. Crutzen, C. Folke, L. Gordon, M. Molina, V. Ramanathan, J. Rockström, M. Scheffer, H. J. Schellnhuber, and U. Svedin. 2011. The Anthropocene: From global change to planetary stewardship. Ambio 40:739–761.

Wootton, J. T., and C. A. Pfister. 2013. Experimental separation of genetic and demographic factors on extinction risk in wild populations. Ecology 94(10):2117–2123.

5 — Species Occurrence in Time and Space

Synthesis and Advancement

> One consequence of a changing world is species don't stay put.
>
> Wiens (2016:91)

Naturalists and scientists aligned with many disciplines have long sought to understand, and then predict, factors underlying the distribution and abundance of species through space and over time. A goal closely aligned with much of this research, which has taken on increasing emphasis in roughly the last 60 years, is how we can use modeling and empirical data to guide management actions that lead to species persistence (i.e., conservation).

In this chapter, our goal is to develop a structured procedure that helps to identify the species and the spatial-temporal contexts that are most relevant for research and management applications. We begin by reviewing the standard terms and approaches, as well as their more recent developments, that have been and continue to be used by ecologists, including concepts of community, species assemblage, metacommunity, and metapopulation. We then argue that the most frequent approaches employed by the majority of ecologists are not substantially advancing knowledge and, as a result, are not providing a basis for management applications that will lead to species conservation. We use the framework that has been established in Chapter 1 to provide specific guidance on how we think scientists can adopt a new framework to accomplish the dual goals of gaining reliable knowledge and promoting species persistence in time and space.

Community

Many scientists, certainly unwittingly, seem to write in a tautological sense by stating that the study of patterns of species co-occurrence is a way to study the structure of ecological communities. Implicit in such an approach is that biotic communities exist as discrete sets of combinations of specific species. Adams (2007) noted that ecologists have long been asking if random assemblages of species form what we call communities, and what role deterministic biotic and abiotic processes play in community formation. As evidenced by the use of this term in a multitude of publications, communities are indeed assumed to exist as discrete, recognizable, biological entities. Thus we must first examine exactly what ecologists and evolutionary biologists have deemed a community. We do this by briefly discussing the importance of a clear and consistent use of terminology, the central issue of establishing the geographic boundaries within which we study ecology, and the specification of which organisms we should include in our studies.

Terminology and the Boundary Problem

As reviewed by Morrison and Hall (2002), ecological terminology has not been not well standardized and has been used inconsistently in the literature. They surveyed definitions of community and found that the co-occurrence of individuals of several species in time and space is common to most. These usually stressed the role of interdependences among the species (populations) under study. Wiens (1989:257–258) concluded that multispecies groups occur in nature, and that we should focus on identifying interactions among these species. He also noted that we cannot hope to understand such groupings of species and the interactions among them if we choose study areas (and therefore species groups) arbitrarily. We cannot use arbitrary and artificial boundaries, because those very boundaries influence what we delineate as a community. The interactions we see among species will reflect our decisions, rather than actual ecological conditions (Morrison and Block 2021).

Fauth et al. (1996) attempted to overcome some of the imprecise language encountered in ecology in order to facilitate clear communication, and make explicit the assumptions that underlie concepts being addressed. They focused on 4 terms they found were often synonymized, ignored, or otherwise misused in the literature of community ecology: community, guild, assemblage, and ensemble. They produced a Venn diagram (Fig. 5.1) in an attempt to operationally define and provide distinct meanings for these often-misapplied terms. They then reviewed the major ways scientists have focused ecological studies, including those that restricted themselves to groups of phylogenetically related species (*Set A*); those that confined their studies to a particular physical area (*Set B*); and those that investigated groups of species that exploit the same resources but considered them without regard to phylogeny or geographic distribution (*Set C*). The intersection of all 3 sets was termed an "ensemble," which was defined as a phylogenetically bounded group of species that

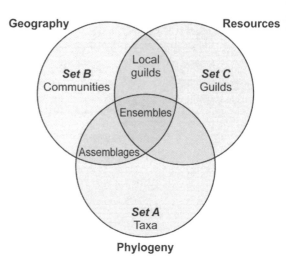

Figure 5.1 Populations under study can be divided into 3 sets, based on phylogeny (*Set A*), geography (*Set B*), and resources (*Set C*). Thus ecological "communities" are merely groups of organisms living in the same place at the same time. By applying this definition of community, the intersections of the 3 sets provide operational definitions for the terms "assemblage," "ensemble," and "guild." Note that no name is applied to the intersection of guild and taxa. Such groups are usually characterized using a compound descriptor defining resources and the taxon involved (e.g., pond-breeding salamanders). (Reproduced from Fauth et al. (1996), with the permission of the University of Chicago Press)

use a similar set of resources within a community. Their use of the word ensemble closely resembles the original definition of "guild" (Root 1967), referring to a set of closely related species that exploit a resource in a common way, such as insect-gleaning canopy warblers. Defining ensembles requires descriptors clarifying the resource, taxon, and geography. Fauth et al. (1996) noted that phylogenetically related groups within a community (the intersection of *Set A* and *Set B*, and therefore a subset of all species in a specified area) are referred to "assemblages," which removes this term from synonymy with community.

Fauth et al. (1996:283) stated that they chose their definition of community "for its simplicity and flexibility; so long as a researcher can place boundaries around her or his study site, a community can be circumscribed easily. The boundaries may be natural (e.g., serpentine soil communities) or arbitrary (all organisms within a 1 m² plot of lawn). The main point is that, in order to constitute a community, the organisms under study must not be restricted further by phylogeny or resource use." They concluded that use of their Venn diagram would enhance communication about the focus of a study and help identify the assumptions hidden in colloquial meanings. Yet this discussion raises the question, "Are communities open or closed systems?"

Looijen and van Andel (1999) noted that 3 primary problems were caused by confusion over the definition of community. First, the term is being used for different entities at different levels of organization, which they said leads to ambiguity. Second, what are usually called communities—namely, populations of different species occurring together in space and time—seldom, if ever, form discrete units in a landscape, but instead gradually blend into one another, making it virtually impossible to draw objective, non-arbitrary boundaries between different communities. This is the boundary problem. Third, communities, as usually defined, are extremely heterogeneous with respect to their species composition, which creates a difficulty, because general theoretical or conceptual constructs are more likely to be valid for homogeneous entities.

Looijen and Arnold (1999:221) went on to define community in a manner they claimed was unambiguous, in which they said that the term "is best used only for groups of coexisting species belonging to a single taxonomic group (class or phylum) in the sense of plants, birds, insects, etc." (They used the term "biocoenosis" for groups of species belonging to different taxonomic groups and forming the biotic components of ecosystems; we will not invoke that term herein.) They specifically stated that a community should not be defined as a group of populations, but rather as the set of individuals of different species occurring in the area of intersection of populations of these species. To them, this definition resolved, "at least in theory," both the boundary

problem and the vexing issue of heterogeneity in community ecology, which they thought would help identify regularities in species co-occurrence patterns in communities. They emphasized that the boundary problem was especially acute in animal ecology and suggested that one solution might be to determine the home ranges of individuals—insofar as the designated species of interest actually do move within specified home ranges—and to map them on top of each other, in order to determine community boundaries. Yet what constitutes a home range is itself fraught with difficulties, given that many organisms shift their use of resources or their locations across seasons.

Looijen and van Andel (1999) thoroughly discussed the boundary problem and presented several diagrams illustrating it when trying to define an ecological community. They proposed that this issue arose from a mismatch with the conceptual factor—namely, that most ecologists at least tacitly define communities as groups of populations occurring together in space and time. In most instances, however, population boundaries of different species in a landscape do not coincide. To illustrate these problems, they developed several Venn diagrams that are reproduced in Box 5.1. What Looijen and van Andel (1999:218) essentially did was formalize the notion that communities are largely arbitrary, or at least vague, and will always represent just parts of multiple species populations: "communities are comprised of only parts, that is, of only certain groups of individuals, of different populations." They discussed various potential problems with communities, including how community composition will change, possibly annually, as population boundaries are altered, due to expansion and contraction. In essence, Looijen and van Andel (1999:218) accepted the realization that the identification of communities will remain in the hands of each researcher, because their definition of it was "the set of individuals of two or more (plant, bird, etc.) species that occur in the intersection of the areas occupied by populations of these species."

Box 5.1. An Illustration of the Boundary Problem in Community Ecology

To illustrate what is wrong, we have constructed the following simplistic but not entirely unrealistic example. Let A1, A2, ... A6 be the areas occupied by populations P1, P2, ... P6 of species S1, S2, ... S6 in a landscape, and assume that within (but not outside) each of these areas relative abundance criteria are met. Assume also that each of the areas overlaps with at least 1 other area, and that the total overlap is such that the intersections of A1, A2, A3, and A4, and of A3, A4, A5, and A6, are non-empty, but that the intersections of A1 and A2, and A5 and A6 are empty (Box Fig. 5.1a). That is, there are 2 distinct but overlapping clusters of populations, the first of species S1, S2, and S3, and the other of species S4, S5, and S6. Suppose that there is a classification system according to which S1, S2, and S3 belong to community type X1; and S4, S5, and S6 belong to another community type, X2. Then, given this typology and assuming that communities are groups of co-occurring populations, one may take the 2 clusters of populations to be communities x1 and x2 of types X1 and X2, respectively. Because the intersection of A3 and A4 is non-empty, however, there is a boundary problem.

Given that communities are groups of populations, it is very difficult, if not impossible, to resolve this problem in an objective, non-arbitrary way—that is, to draw an objective, non-arbitrary line separating x1 from x2. We might use the lower (southern) boundary of P3, but then parts of P4, P5, and P6 would come to belong to x1, making it no longer a community of type X1. A similar problem occurs if we use the upper (northern) boundary of P4. To solve this difficulty we might take the upper bound-

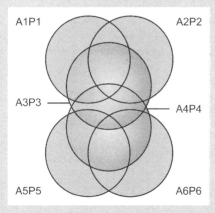

Box Figure 5.1a Schematic representation of the areas *A1* through *A6*, occupied by populations *P1* through *P6* of species *S1* through *S6*, respectively. There are 2 more-or-less distinct clusters—the first of *P1*, *P2*, and *P3*; and the other of *P4*, *P5*, and *P6*—that may be seen as 2 different communities. The areas they occupy are defined as the conjunctions of *A1*, *A2*, and *A3*, and of *A4*, *A5*, and *A6*, respectively. Because *A3 ∩ A4* is non-empty, there is a boundary problem. (Reproduced from Looijen and van Andel (1999), with the permission of Elsevier Science and Technology Journals)

ary of P4 to also be the lower boundary of x1, and the lower boundary of P3 to also be the upper boundary of x2. This would have the advantage of x1 and x2 being "pure" communities of types X1 and X2, respectively. The disadvantage, however, would be that the interjacent area (the intersection of A3 and A4) would not be taken into account. To solve—or rather to diminish—that problem, we could also use the lower boundaries of P1 and P2 and the upper boundaries of P5 and P6 as the lower and upper boundaries, respectively, of 2 different communities, x3 and x4. In that case, these communities would not be of types X1 and X2, but of, say, types X3 and X4, respectively, the former being comprised of species S1, S2, S3, and S4, and the latter being comprised of species S3, S4, S5, and S6. Moreover, there would still be an interjacent area left out—that is, the area in between x3 and x4, containing parts of P3 and P4. To resolve this latter problem, we could define this interjacent area as a transition zone or boundary area separating x3 from x4. Alternatively, we could define the intersection of A3 and A4 as a boundary area separating x1 from x2. Defining transition zones or boundary areas has been a "solution" to the boundary problem. In addition, the observation that the species composition in such boundary areas is always more or always less different from that in either of the "main" communities they "separate," has led to the creation of so-called boundary communities.

Each of the above options is more-or-less unsatisfactory, and it is probably for this reason that most ecologists believe that the definition of communities must always be somewhat arbitrary. Moreover, to define (and classify) boundary communities does not really to solve the boundary problem, but instead accepts it as being irresolvable. This is not the only difficulty, however. If population boundaries do not coincide, then lumping groups of populations into communities will always create communities that are spatially heterogeneous in their species composition.

Consider again the above example. This time, let us look at the case where a boundary community (say x6) in the intersection of A3 and A4 separates communities x1 and x2. Each of these communities is heterogeneous, in the sense that the species composition in 1 part of the total area it occupies is different from that in another part (Box Fig. 5.1b). Within the boundaries of x1, for example, the species composition in area 2 (S1, S2) is different from that in, say, area 5 (S1, S2, S3) or area 6 (S2, S3). Similarly, within x2, the species composition in area 15 (S4, S5, S6) is different from that in, say, area 14 (S4, S5) or area 16 (S4, S6). Each of the intersection areas in Box Figure 5.1b, including those within the boundaries of x6, has a different species composition.

As mentioned before, the heterogeneity of communities may be a reason for the present lack of (well-confirmed or corroborated) general laws and theories about communities. Unfortunately, the present classification systems tend to aggravate the problem of developing such laws and theories. The major criticism of these systems is that communities, as defined by them, are hardly ever found in nature, at least not in their "pure," "ideal," or "typical" form. Actual communities in the field (phytocoenoses) are almost always "imperfect reflections" of the ideal types (phytocoena). Either some species are lacking that, according

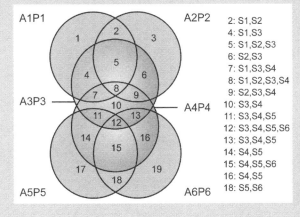

2: S1,S2
4: S1,S3
5: S1,S2,S3
6: S2,S3
7: S1,S3,S4
8: S1,S2,S3,S4
9: S2,S3,S4
10: S3,S4
11: S3,S4,S5
12: S3,S4,S5,S6
13: S3,S4,S5
14: S4,S5
15: S4,S5,S6
16: S4,S5
18: S5,S6

Box Figure 5.1b   The same example as in Box Figure 5.1a. Here, each of the numbered areas has a different species composition and, therefore, a different (type of) community. Areas 1, 3, 17, and 19 each contain individuals of only a single species. Thus they cannot be said to be occupied by a (multispecies) community. (Reproduced from Looijen and van Andel (1999), with the permission of Elsevier Science and Technology Journals) to the system, should be present; or others are present that, according to the system, should be absent; or both. The chief cause is the fact that community types (phytocoena) are abstractions from sets of concrete communities (phytocoenoses) whose members are often (1) heterogeneous, and (2) more or less variable and, hence, different from one another in terms of species composition (though perhaps not as different as they are from members of other sets). That is, both the sets and their members are often heterogeneous. Both forms of heterogeneity disappear, however, in the abstract types (which are more-or-less fixed species combinations). As a result, the communities we observe in the field generally deviate considerably from the ideal types. This problem would not arise if community types were abstracted from sets of communities that are identical and homogenous in terms of species composition.

For various reasons, ecologists have been unable to solve either the boundary problem or the problem of heterogeneity, and they seem to accept them as being insoluble. This is odd, because in our view, there is a simple answer to both problems. To arrive at this, recall that both problems result from the fact that communities are commonly defined as groups of co-occurring populations, whereas, in actuality, different populations seldom co-occur in exactly the same area. What we have in mind here, and throughout our chapters, are local populations of different species, not metapopulations; the metapopulation concept is irrelevant to this book. As a result, there will always be populations that range over communities, no matter how the community boundaries are drawn. Thus communities will always be heterogeneous, and community boundaries will always be vague. If populations range over

communities, however, then communities cannot be groups of populations! Rather, populations are cut up into some parts belonging to a particular community and other parts belonging to another community. There is no way of defining 2 or more communities in Box Figures 5.1a or 5.1b without breaking up at least some of the populations. In other words, communities are comprised of parts—that is, of only certain groups of individuals—of different populations.

To arrive at a more precise definition, let us take another look at Box Figure 5.1b. Strictly speaking, the set of individuals in the transition zone of x1 and x2 (the intersection of A3 and A4) cannot be regarded as a community in the sense of either type X1 or type X2, since these types are defined as combinations of species S1, S2, and S3, and species S4, S5, and S6, respectively, whereas the species composition in the transition zone is S1, S2, S3, S4, S5, and S6. This is the reasoning behind the introduction of a boundary community. The solution to the boundary problem, however, is not to define boundary communities, but to take the rationale behind them more seriously. This rationale applies not only to the transition zone, but also to each of the intersection areas in Box Figure 5.1b. That is, strictly speaking, a community x1 of species S1, S2, and S3 occurs only in the intersection of A1, A2, and A3 (area 5), and it is not true that a community consisting of these species is found in any of the other (intersection) areas. Similarly, a community x2 composed of species S4, S5, and S6 occurs only in area 15, and it is not true that such a community occurs in any of the other intersection areas.

We can speak of a community of 2 species, A and B, only in the intersection of the areas occupied by populations of these species, where individuals of both A and B occur, not in areas where individuals of only A or B appear, nor in areas where individuals of both A and B and of another species C occur. This implies, however, that communities are not to be defined as groups of co-occurring populations, but as groups of co-occurring individuals of different populations. More precisely, then, a (plant, bird, etc.) community may be defined as the set of individuals of 2 or more (plant, bird, etc.) species that occur in the intersection of the areas occupied by populations of these species. The significance of the word "the" starting this definition should be noticed, as it is so much more definite than the words "an" or "any" in former community definitions (such as "C1–C9"). We shall denote our definition as CI (community of individuals).

Source: Looijen and van Andel (1999)

Looijen and van Andel (1999) seemed primarily concerned with using a population level as the basis for community identification, because population boundaries can change often, meaning that the designated community would be very unstable. We agree with their assessment but want to point out that this instability is really a result of their definition of community, not an inherent ecological problem. Rather, fluctuating population boundaries are at the core of understanding how the distribution, abundance, and productivity of species (or subspecies and ecotypes) change in time and space. From the standpoint of population persistence (and, thus, conservation of a species), knowing what is occurring, such as along the periphery of a population, is essential for development of potential management applications (e.g., see Chapter 6). We fail to understand how treating a community as a Venn diagram of populations can, in the long run, represent meaningful ecological processes.

Vellend (2010) noted that there has been considerable debate in ecology concerning the degree to which ecological communities—particularly faunal ones—are sufficiently coherent enough entities to be considered appropriate objects of study. Vellend (2010:187) defined community as "a group of organisms representing multiple species living in a specified place and time." He further stated that he chose this definition because it implicitly embraced all scales of space and time. Here again, we find a definition of community that is extremely vague.

Vellend (2010) did, however, develop what we think is a useful and straightforward way to conceptualize the processes that are ultimately responsible for the patterns of species distribution and abundance that we observe. The 4 primary processes identified by him were speciation, drift, selection, and dispersal (Fig. 5.2). Hence the role of research is to determine the relative influences of these 4 processes through space and over time. In Chapter 1,

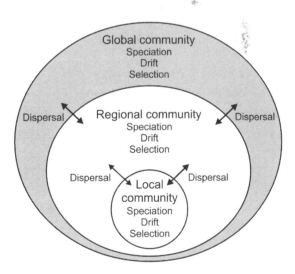

Figure 5.2   A diagrammatic representation of how speciation, drift, selection, and dispersal interact to determine community structure across spatial scales. The delineation of discrete spatial scales is, however, arbitrary and is used only for clarity in this presentation. (Reproduced from Vellend (2010), with the permission of the University of Chicago Press)

we have discussed on the central importance of identifying populations by delineating the boundaries of a selective environment—that is, the evolutionary perspective of how animals became adapted to a set of environmental conditions (Millstein 2014). This is because populations are a unit of evolution, and they determine the boundaries of their selective environment. Thus these boundaries are determined by the spatial extent of the population.

Food Webs and Community Structure

The study of food webs is plagued by the same boundary problem that permeates studies framed as focusing on animal communities. In his review of community food web studies, Cohen (1990) defined a community as whatever animals lived in a habitat that some ecologist wanted to study. Although he did not use "habitat" in the species-specific sense, he clearly meant the species that occurred over some vegetation or other identifiable area, based on human perception. Cohen (1990) went on to state that once the physical boundaries of a habitat are defined, one can study flows of matter and energy across and within boundaries. We do not disagree with his comments on food webs, and it is not necessary for our thesis to review food web theory and study. As noted below, we think that the examination of food webs is obviously a central part of the study of animal ecology. Our criticism of how ecologists approach "community structure" through research on behavioral interactions, however, applies equally to how ecologists approach community structure through food webs and trophic structures. As reviewed by Montoya and Sole (2003), much of the work on food webs has concerned how patterns emerge from the population dynamics of interacting species—that is, how communities are structured. As they noted, the literature on assembly dynamics (for instance, more recently, Hein and Gillooly 2011) has tended to center on the role of competition, with much less work focused on trophic structure, speciation, and drift.

As initially developed by May (1973) and reviewed by Rooney and McCann (2012), increasing species diversity has tended to destabilize community dynamics in the absence of a stable food web structure. Rooney and McCann (2012) noted that it was the arrangement of strong and weak interactions in food webs that drove diversity, stability, and, thus, community structure. They concluded that the history of food web ecology had been focused on a diversity-stability domain and a structure-stability domain, with little emphasis on the relationship between diversity and food web structure (i.e., the diversity-structure domain). Here again, we think it is easy to conclude that the knowledge of species' interactions, dynamics, biodiversity, and all such related matters will increase substantially—and allow meaningful comparisons across time and space—when we are working within known biological populations.

More recently, the many ways that organisms interact has been lumped under the general term "ecological networks," which focus on food webs and trophic-level relationships. Such research has been broadly categorized into 3 types of networks: traditional food webs, host-parasitoid webs, and mutualistic webs (Ings et al. 2009). These authors noted that in community studies, the focus has been on individuals (the nodes in the network) that compose species populations, with the links connecting them reflecting population effects. We agree in principle with this statement but note that a proper application of any form of network analysis requires a rigorous determination of the populations.

Although Schmitz et al. (2017) did not explicitly discuss using a biological population as a focus of study, they clearly recognized the fundamental role animal populations play in predator-prey interactions and habitat domains within the context of community ecology (where they defined "habitat domain" as the spatial extent of habitat over which individuals move in the course of their foraging). For example, they depicted various scenarios (Fig. 5.3) where exploitative competition could arise when

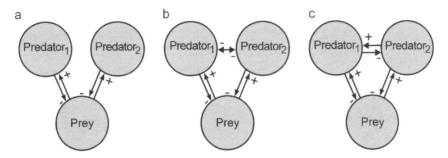

Figure 5.3 A depiction of exploitative competition (*a*), interference competition (*b*), and intraguild predation (*c*) in a spatial context, using the habitat domain concept. This figure predicts a priori conditions needed for different natures of multiple predator-prey interactions to emerge, and it illustrates that spatially, there may be more than a single type of emergent effect for a particular food web.

prey have a large habitat domain and predators have small overlapping domains, or when predators have large overlapping domains that overlay the small domains of prey (*a*). They also postulated that predator interference could occur when both predator species and prey overlapped spatially, as well as when either all predators and all prey have large habitat domains or they all have small habitat domains (*b*). Lastly, they predicted that intraguild predation could occur when predator species have small overlapping domains and prey have large habitat domains (*c*). Schmitz et al. (2017) recognized that their analyses were focused on individual population members but posited that generating a population-level habitat domain could be accomplished by aggregating the behavior-specific utilization distribution of each population member (see also Miller et al. 2014). If community ecology is to be a truly integrative science, as Schmitz et al. (2017) encouraged, all of the plausible scenarios they provided highlight the need to expand analyses out to the population level to fully understand the interactions between organisms and their biotic and abiotic environments, as well as how species are distributed and persist over time.

Here again we remind readers that it is not our intent to thoroughly review and debate the processes driving species occurrence, but rather to develop a reasonable framework under which such mechanisms can be empirically identified. It seems apparent enough to us, however, that population remains

the key focus of most descriptions of community. The authors of most papers almost always seem to snip and cut biological populations into overlapping parts to define various communities. Community ecology was developed in large part through studies of individual animals interacting at the same place and time. We are not criticizing this type of research in any way. What we are criticizing is the often tortuous manner in which ecologists have sought to justify their largely arbitrary identification of "community" and attach to it some ecological—and often evolutionary—meaning.

Because biological population seems substantially more empirical than the concept of community, we think it makes much more sense to focus on populations and their related constructs (e.g., subspecies, ecotype). As part of his review of the history of community ecology, Wiens (2016:227) asked, "Are bird communities real? Do they exist as functional biological entities, or are they only constructs in the minds of ecologists?" Although he acknowledged the value of studying closely related species, he highlighted the perils of arbitrarily truncating functional webs, especially when dealing with unrelated species. He then concluded, "If 'communities' are to be a target of conservation efforts, they should exist in something other than our minds and publications." Wiens (2016) went on to recommend that we should focus more on populations and environments in relation to defined and measured resource bases. He also noted that if organisms are not just haphazard assemblages, they must arise from various biotic process. These would then be expressed through individuals and populations.

Species Assembly

In this and the following section, we first briefly review the development of species assembly as it relates to what we then discuss: the generally accepted concepts of community and metacommunity. As succinctly reviewed by Adams (2007; see also Weiher et al. 2011), a contentious debate ensued following Diamond's (1975) development of assembly rules, which sought to explain the underlying mechanism for non-random occurrences of species across space in an island archipelago. In addition to Diamond's (1975) focus on competition as a driving force in assembling groups of species, the proper choice of and role for developing null-model communities in analyzing species occurrences has long been debated (e.g., Connor and Simberloff 1979; Simberloff et al. 1999; Gotelli 2000). In summary, species assembly was based initially on concepts presented by Diamond and other colleagues (see Cody and Diamond 1975), who suggested that competition was the driving influence. Subsequent work has indicated that, instead, biotic and abiotic factors operate stochastically and deterministically to influence the suite of species occupying an area for an indeterminate amount of time. Assemblages are rarely at equilibrium, because both endogenous and exogenous factors influencing species distributions are constantly at work, enabling species to move into, within, and out of the system. A modern example of this is the emergence of "novel ecosystems," which are combinations of species that have never before occurred together but are formed when such species have differentially shifted their range distributions in response to climate change (Clement and Standish 2018).

Wiens (1981, 1989) provided a history of the development of the concepts of community and species

assembly. Wiens (1989) concluded that we require more than inventories of presence-absence of species at particular times. We need to quantify habitat relationships and ecological attributes, population densities, resource levels, and the abundances of predators and competitors. He emphasized that his criticism of community studies (or investigations in general) must be taken in context: we, as scientists, learn by doing. Wiens (1989:17) noted that few of the people who figured prominently in his book "would unhesitatingly embrace statements they made 20 or 10 or perhaps even 5 years ago." As Wiens (1981) reminded those who approach the study of pattern and process, we need to be careful to avoid adhering to a paradigm, such that our methodology becomes "normal science" (sensu Kuhn 1970). This is because the research tradition established by the paradigms directs the perspectives and procedures to be following by investigators.

We take care to remind readers that the very people we are criticizing are the same ones who are largely responsible for advancing knowledge. Their work is not obscure, because they were at the leading edge of the discipline! Pioneers in community ecology have fostered a tremendous amount of research. One would hope, however, that we can also learn the weaknesses of any approach and then move forward, building off the foundation that has been constructed. For instance, in Chapter 7 we incorporate the species assembly concept into our recommendations for approaching investigations and advancing the study of animal ecology.

As reviewed by Weiher et al. (2011), ecologists have tended to take the view that community assembly is broader than just a happenstance being mediated by competitive versus non-competitive processes and is, thus, a more general phenomenon. They concluded that community assembly has been viewed as a broad and well-accepted concept, and that rules per se are now rarely mentioned. Thus it is the process determining the non-random assembly of species that we are focusing on in this book, rather than arguing for a specific set of rules. When non-random patterns of assembly are observed, interest in them obviously becomes the identification of the processes underlying those patterns. Are these patterns due to deterministic processes, including interspecific competition, resource exploitation, environmental heterogeneity, or other effects? Or are the processes stochastic? As noted by Adams (2007), although the identification of non-random patterns can indicate that a particular process is prevalent across a set of communities, such an attribution must be done with care. Manipulative experiments (see Chapter 1), especially those conducted in the field, can help elucidate and provide strong evidence for the processes in operation under different conditions.

Community ecologists have often invoked a niche-based, pool filter–subset concept of community assembly, in which local assemblages are viewed as subsets of a regional species pool that is delimited by a set of filters. These have been characterized as abiotic fundamental-niche filters that determine whether a species has the requisite traits to colonize, become established, and persist in a given habitat; and as a set of biotic realized-niche filters that are imposed by the interactive milieu of competitors, mutualists, and consumers (Weiher et al. 2011). If a species cannot survive in a particular area, then their previously inhabited area is no longer habitat for that species.

A niche-based approach to assembly does not mean that the specific combination of species present is static, or even highly predictable. Communities can converge, based on functional parameters where different subsets of the regional pool can develop, following historical dispersals. Thus there should not be an expectation of equilibrium—that is, where a specific combination of species persists unchanged over time—in any local community, especially given that, without substantial external input (e.g., a natural disturbance or human-generated management actions, such as regular cutting or burning), ecological succession will take place. Thus assembly is an active, ongoing process of filtering and sorting, with assemblages persisting for various lengths of time. Because some communities will last longer than others, the context underlying studies being conducting on various aspects of animal ecology, behavior, and genetics must be continually re-evaluated. Additionally, neutral processes must be simultaneously considered, because they can occur at the same time as niche-based processes. Weiher et al. (2011) thought it was more appropriate to think of niche-based processes as affecting the likelihood of particular mixtures, as opposed to determining them.

As reviewed by Weiher et al. (2011), neutral theory means that communities are no longer considered to be random assemblages. Rather, they are neutrally or stochastically put together by dispersal, ecological adaptations, and historical inertia. Myers and Harms (2009) called them "dispersal-assembly models." Neutral theory requires us to distinguish an assembly, which may be deterministic and induced from niche-based relationships, from a neutral assembly, which entails no interspecific causal interactions. This theory applies where stochastic events (e.g., climatic variability) strongly affect community assembly, due to deterministic, niche-based processes that may be unpredictable. Thus assembly may appear to be random. Under niche assembly, the same stochastic processes occur, but they are considered to be of considerably less importance, because species are not equivalent. Weiher et al. (2011) concluded that the neutral-versus-niche debate has been unproductive, whereas a synthesis of these concepts would be clearly more valuable in advancing our understanding of species assembly (see also Leibold and McPeek 2006). The synthesis of these concepts into a unified model of assembly is advancing, with a goal of improving our empirical understanding of where communities fall along the niche assembly–neutral (dispersal) assembly continuum (Myers and Harms 2009). Vellend (2010) argued that species diversity—an outcome of assembly—is determined by neutral and stochastic processes, selection (niche processes), dispersal (a combination of stochastic and trait-dependent processes), and speciation (which shapes the species pool). Our goal in this chapter is to provide guidance on how to determine the relative strength each of these processes exhibit in assembly. Vellend (2010) also considered speciation (and, to a degree, subspeciation) to be relevant within a species pool itself, because local (community-level) assemblages are influenced by the broader processes that have shaped the regional species pool.

Although Agrawal et al. (2007), in their review, noted that community ecologists have moved from a primarily competition-based study approach to one that incorporates predation, mutualism, and parasitism, as well as disease and other interactions. Agrawal et al. (2007) also recognized the importance of outcomes of interactions, indirect effects, trait-mediated interactions, and intraspecific genetic variation. They concluded, however, that advances in community ecology have remained limited by, among other factors, (1) a failure to adequately consider how biotic and abiotic contexts shape the strength of species interactions; (2) the degree to which the distribution and abundance of a given species are influenced by interspecific interactions; (3) how biotic and abiotic factors interact and vary in magnitude over time or through space; (4) how a variation in the abundance of particular species influences that of the species with which they interact; (5) the scale dependence of feedbacks between community interactions and environmental conditions; and (6) the mechanisms driving the relationship between species diversity within communities and genetic diversity within populations. We certainly agree with their summary and highlight their focus on space and time, interspecific interactions, the issue of population-level genetics, and the overall biotic-abiotic context established for investigation. In other words, Agrawal et al. (2007) seemed to be calling for a recognition of how species interact with each other and their environment across space and time, along with population-level considerations, yet they did not speak to the inherent problem of setting artificial boundaries around these parameters, in order to establish a community to be studied.

Metacommunity

Leibold et al. (2004) noted that much of community theory had assumed a single and ecologically small scale (geographic extent), termed "local communities," that are closed and isolated. Within local communities, populations were assumed to interact directly in various ways (e.g., affecting each other's birth and death rates). These authors understood, however, that because ecological processes involving other species occurred across other scales of spatial extent, a spatially broader context was needed in community ecology. They then drew the conclusion that species interactions took place in a network of local communities at larger scales, which they referred to as "metacommunities" (see also Sokol et al. 2017). Leibold et al. (2004:602) defined a metacommunity "as a set of local communities that are linked by dispersal of multiple potentially interacting species."

They also noted that metacommunity thinking had already led to its own terminology, which was strongly influenced by the study of metapopulations. Leibold et al. (2004:Table 1) defined "population" as "all individuals of a single species within a habitat patch." They then depicted "patch" as "a discrete area of habitat. Patches have variously been defined as microsites or localities. In this chapter we use the term analogously to localities, which are capable of holding populations or communities." It seems that they were equating habitat with a physical area that could contain multiple species, which runs counter to the (species-specific) definition of habitat, but which is certainly present in the literature (see Chapter 2). This definition is also related to the confusion between vegetation patches and habitat patches (see Chapter 3).

From our discussion in Chapter 1, readers should quickly realize that we take substantial exception to this terminology. We must assume that Leibold et al. (2004) were not concerned with a biological population, because such an entity seldom occurs within a patch (unless that patch is extremely large, or the focal species has a small body size and is sedentary or highly locally endemic). Definitions provided by Leibold et al. (2004:Table 1) for microsite (capable of holding a single individual), locality (encompassing multiple microsites and capable of holding a local community), and region (containing multiple localities and capable of supporting a metacommunity) are welcomed, but they have revealed that the metacommunity concept, like its component parts (communities), have both been based on an observer-defined preference or understanding of scale, not on what the organisms (i.e., individual animals that are assembled into biological populations) have been telling them. Thus, we note that Leibold et al. (2004) allowed the notion that community was something that could be variously defined by the observers.

Leibold et al. (2004) understood that an application of the metacommunities concept to empirical situations was complex because (1) local communities rarely had discrete boundaries, and (2) different species might respond to processes at different scales. These comments by Leibold et al. (2004) highlight the fundamental problem inherent in the basic concept of community in ecology, which is substantially compounded when one tries to link multiple communities into some type of network of real or potentially interacting species in nature. We again see the boundary problem (which, as we noted above, is only a problem because the community concept makes it so) as a paramount issue. It does not provide a means by which to discern how species interact with both other species and the environment across space.

Leibold et al. (2004:Table 1) thoroughly defined terms and identified the multiple models or paradigms attached to the concepts of community and metacommunity, such as classic (Levins 1968) metapopulation and source-sink systems, as well as those attached to metacommunity concepts, such as patch dynamics perspectives and species-sorting per perspective (see Leibold et al. 2004:Fig. 1). We do not think that these models or paradigms should be ignored. They represent a wealth of potential hypotheses regarding the patterns that emerge from empirical field studies, and they are very useful in study design. As we have developed herein, what we are suggesting is that the overriding paradigm needs to centrally focus on sampling from the biological population.

Leibold et al. (2004:608) concluded that all 4 metacommunity paradigms captured aspects of metacommunity dynamics, but added, "It is unlikely that all the species that interact in a given set of real metacommunities will uniformly conform to any one of these perspectives. Instead, it is likely that each of these sets of processes will play interactive roles in structuring real metacommunities." We are admittedly troubled by their reference to "real" in describing the concept of a metacommunity. We are not criticizing the various potential dynamic scenarios offered by Leibold et al. (2004), as there is a large body of empirical evidence dealing with local species interactions, dispersals across time and space, and so forth. What we are critiquing is a research focus on broadening what we consider the very abstract concept called a community.

Metapopulations

Our concern with the metacommunity approach rather naturally leads to a consideration of the metapopulation concept. Despite our criticism of the value of metacommunity approaches, we—along with probably all ecologists—understand that there is a spatial element in ecology, such as the causes and effect of disjunct distributions of species, and related differences in genetics and vital rates. The formal concept of a population presupposes the presence of metapopulations. The metapopulation concept arose in an attempt to understand the relationships within and between population segments that could be connected, at least occasionally, by dispersing (emigrating and immigrating) individuals. Especially because of the perceived application of metapopulation theory to conservation goals, this area of research has gained substantial attention, both theoretically and empirically. In Chapter 1, we have discussed the metapopulation concept in the context of its role in potential speciation; we review and provide guidance on population delineation later in this chapter.

Here, we are interested in the distribution of individuals belonging to a particular species, as well as whether some or all of those individuals can be empirically classified as populations (or subpopulations within a metapopulation). We care about this because our ultimate goal is to understand what drives the persistence of a species, enabling us to offer management guidance that ultimately provides for this. In this chapter, we are not trying to compare various statistical analyses or models. Rather, we outline a framework that can be used to approach this ultimate question, as many current and certainly future studies and models will be used to analyze appropriate data. As Hanski (2004) had concluded, we know very little about the spatial population structure of the majority of species, a statement that still holds true many years later.

Despite the considerable attention given to metapopulation theory, most investigators have carried along some unfortunate misuses of terminology and, in doing so, have unintentionally weakened their efforts. Most notable is a common attempt to divide the environment into habitat area–based subcomponents, usually termed a "patch" or "habitat patch." By definition, a habitat patch is nonsensical, because it means there is an identifiable habitat that is separate from the animal (Morrison and Block 2021). Recall that habitat is a species-specific concept and must not be used synonymously with a physical area per se. Thus a major problem with an application of the metapopulation concept is a tendency to ignore the subpopulation structure—which, along with dispersal, is the fundamental component of a metapopulation—and instead focus on a physical, observer-defined area called a patch. There can certainly be cases where a subpopulation is shown to be contained within a specific physical area, but this island-like scenario seldom exists. Rather, subpopulations are usually distributed across an environment that, while perhaps being more heterogeneous within

subpopulations, as opposed to between them, is nevertheless heterogeneous, or patchy. Unfortunately, focusing on a patch largely takes us back to the boundary problem of community and metacommunity. Further, other terms and concepts related to metapopulation theory that may be problematic include "unsuitable habitat" (an oxymoron) and "matrix" (conditions assumed to be unsuitable for breeding but suffice for successful dispersal between and among subpopulations).

Various researchers have realized, usually without explicitly identifying the patch terminology issue, that there were problems inherent in the way typical metapopulation studies were being conducted. For example, Pellet et al. (2007) noted that the locations, sizes, and occupancy states (occupied or unoccupied) of patches in a metapopulation could be determined on the basis of field surveys. They added, however, that gathering data on density, vital rates, immigration, and emigration usually was difficult. Their research showed that area alone often was an inadequate surrogate measure of population size and, therefore, was less effective than population size for predicting the extinction of patches in a metapopulation. They also found that connectivity, which is a traditional metric for isolation, performed poorly in predicting probabilities of patch colonization. They thus encouraged researchers to gather data on other parameters, including habitat quality, in metapopulation models to improve the predictability of the long-term persistence of species. Pellet et al. (2007) warned that conservation planners should be aware of the potential limitations of metapopulation models based only on patch area and isolation. A central point to remember is that a fragmented population is not necessarily a metapopulation, and that, for management purposes, metapopulations differ considerably from populations with disjunct distributions (Esler 2000). We emphasize again (see Chapter 1) that a fragmented or disjunct group of animals does not necessarily compose a subpopulation, nor do sets of such groups necessarily make up a metapopulation.

As reviewed by Esler (2000), metapopulation theory has focused almost exclusively on non-migratory species. As he noted, however, population structure (i.e., a single population, multiple isolated populations, a metapopulation) and interactions might vary across the annual cycle of a species, which could be especially pronounced in migratory species (discussed later in this chapter). Additionally, a population could be demographically panmictic (all of them interbreeding, so there is no metapopulation struc-ture) for part of the year, yet be structured as a metapopulation during another period. Esler (2000) focused on the identification of subpopulations; his only reference to a patch was in an island situation. Expanding the metapopulation concept to include migratory species will present substantial difficulties in design and field application, but it will also provide a potentially fruitful way to expand our understanding of population dynamics and persistence in a holistic manner (see Box 5.2).

Box 5.2. Spatial and Temporal Dynamics of Migratory Birds between Breeding and Wintering Areas

Although geography is not the only determinant of demographic independence in migratory birds, geographic isolation must occur at some critical stage in the annual cycle for demographically independent subpopulations to exist. In general, 4 scenarios of spatial and temporal dynamics of migratory birds during breeding and winter are possible (Box Fig. 5.2). These are obviously simplifications of the annual cycle; they do not include staging, molting, or migration areas, for example. Yet most of the temporal and behavioral factors that could lead to subpopulation independence occur during breeding and wintering, and these generalizations can be extended to other periods.

The situation when winter panmixia and distinct breeding areas occur (A) could be considered a metapopulation if there is some behavioral mechanism that results in demographically independent breeding subpopulations—that is, if these subpopulations have independent extinction probabilities. This is clearly different from the typical non-migratory metapopulation, in that

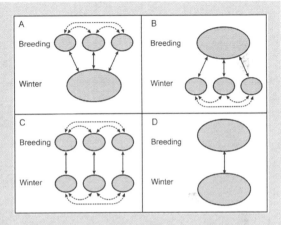

Box Figure 5.2   Four scenarios of the spatial and temporal dynamics of migratory birds between their breeding and wintering areas. *Ovals* represent geographically distinct groups, *solid arrows* represent migration, and *dashed arrows* represent dispersal. (Reproduced from Esler (2000), with the permission of John Wiley & Sons)

individuals co-occur during winter. Co-occurrence, however, does not preclude the demographic independence of breeding subpopulations, in which case metapopulation theory can apply.

The situation when distinct winter areas and breeding-area panmixia occur (B) is similar to the previous one and could function as a metapopulation if behavioral mechanisms exist that create demographically independent wintering subpopulations. In other words, even though all individuals share a breeding area, demographically independent subpopulations could exist on wintering areas if, for example, philopatry there is high. The applicability of metapopulation models, moreover, is not limited to the breeding season. Subpopulation independence, whether on breeding or wintering areas, can have important implications for metapopulation dynamics and persistence.

For migratory birds, distinct winter and breeding areas (C) is the situation that most closely parallels the conventional non-migratory metapopulation model. Linked breeding and wintering areas could represent distinct subpopulations, with an occasional exchange of individuals through dispersal at any part of the annual cycle. If breeding and wintering areas are not linked—if, for example, birds in a wintering area come from several breeding areas (or vice versa)—the metapopulation may function as in the previous 2 examples, or the population may be demographically panmictic and thus not qualify as a metapopulation at all.

The situation when winter and breeding-ground panmixia occurs (D) is analogous to a single panmictic non-migratory subpopulation. There is no mechanism for subpopulation demographic independence, so metapopulation theory cannot apply.

Source: Esler (2000)

Genetic diversity and the related issue of effective population size (the proportion or number of individuals in a population that are actually breeding) are frequently referenced in metapopulation studies. This is because the effective population sizes of animals within a metapopulation structure are generally thought to be much smaller than the number of mature individuals. This can either reduce genetic diversity or result in stochastic population crashes, especially in small-sized and isolated subpopulations. Because of the central importance of population size and dispersal in the persistence of a species, we see yet another reason why a focus on a physical area, without a direct link to the actual distribution of the subpopulation in question, is inappropriate.

Determining the population structure for migratory species is hampered by difficulties in identifying demographically independent subpopulations. As noted above, a fragmented or disjunct group of animals does not necessarily compose a subpopulation. Additionally, co-occurring migratory individuals of a species might not be part of the same demographically panmictic group (Esler 2000).

One example is work conducted across the entire breeding range of federally endangered golden-cheeked warblers (*Setophaga chrysoparia*). An initial federal listing was based, in part, on assumptions, including a metapopulation structure and concerns about a loss of genetic diversity. Subsequent study showed that the species exhibited no subpopulation structure (Morrison et al. 2012). The species does, however, occupy what many would call a very patchy environment, in that it occurs primarily in mature oak-juniper (*Quercus-Juniperus*) woodlands across the central part of Texas (Fig. 5.4). Research conducted across the breeding range of this warbler species showed that occupancy was driven primarily by their geographic location within this range and the size and location of oak-juniper patches (Collier et al. 2012; Mathewson et al. 2012).

Ideas for Population Ecology

Armstrong (2005) discussed blending what he called the "habitat paradigm" into the "metapopulation paradigm," rather than treating each of them as largely

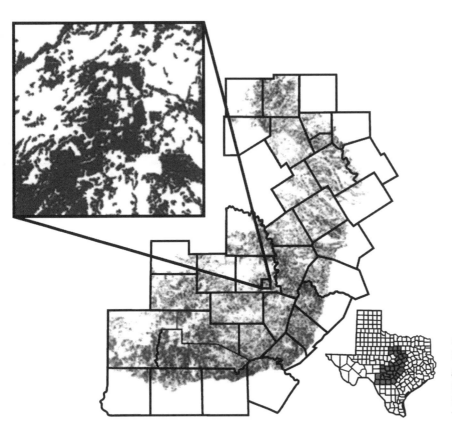

Figure 5.4 The probability of occupancy by golden-cheeked warblers (*Setophaga chrysoparia*) on their breeding range in Texas.

separate methodologies. He was concerned that approaches based solely on the metapopulation or habitat paradigms would lead to only limited management recommendations. Those resulting from a metapopulation approach often suggested that corridors or stepping stones be established to facilitate movement, and that reintroductions (or translocations) of individuals should take place to recolonize or bolster certain subpopulations (see Chapter 3). On the other hand, recommendations stemming from a habitat-based approach focused primarily on how to alter habitat conditions to improve habitat quality. Armstrong (2005) further explained that metapopulation and habitat factors were confounded, as habitat quality can be low because of small area size, edge effects, and various other biotic and abiotic factors. Additionally, isolation could be correlated with habitat quality, because isolated habitat remnants were usually abandoned because of their poor quality (e.g., limited food resources, presence of a predator, Allee

effects), rather than simply because of demographic stochasticity. He also noted that information on vital rates was seldom gathered, and assumptions on habitat quality were made indirectly, using surrogates (e.g., abundance). Moreover, regardless of the paradigm being followed, information on vital rates on a site-specific basis was the only sure way to address habitat quality. As summarized by Hanski (2004), metapopulation models have not "ignored" demographic processes within local populations, but these processes have not been explicitly modeled. He also noted that for some purposes, focusing at the metapopulation level was sufficient, while for others it was not (see also Ovaskainen and Hanski 2003). For example, Vögeli et al. (2010) found that metapopulation modeling that included habitat quality has been shown to strongly improve estimates of species occurrence, colonization, or extinction patterns.

We certainly agree with Armstrong's (2005) goal to enhance how environments are analyzed within a broad spatial scale, including the concept of meta-populations. There is nothing incorrect per se about just looking at occupancy; investigators are always expected to provide a rationale for the approach they took in a study. Nevertheless, for our purposes here, we are interested in investigating how we can substantially expand the information we gain on biological populations, which naturally requires examining the distribution of relevant populations in space and time, along with the factors driving performance (i.e., vital rates). A metapopulation focus that ignores vital rates will inevitably conflate processes associated with population fragmentation with those of habitat loss and habitat degradation, as the latter 2 would be occurring regardless of spatial configuration. This is yet another reminder of why we do not think that the concepts of community and metacommunity have the heuristic power needed to jump-start our knowledge of animal ecology and its applications to management to a much higher level.

We also agree with another point made by Armstrong (2005): collecting data on vital rates can be labor intensive and require multiple years to accomplish, especially over broad spatial extents. But we fail to see how we can substantially expand our knowledge of animal distributions, abundance, and, ultimately, persistence without gathering reasonable data on population structure and vital rates.

In any case, Armstrong offered 3 solutions for improving research.

1. Where previous research has been conducted on a species, 1 or 2 relatively simple parameters (e.g., nest success) might be found to accurately predict the population growth rate. These can be used as reliable measures of habitat quality, to be integrated with metapopulation dynamics.
2. Where little is known about a species, thus necessitating a shotgun approach to habitat analyses, attempting to integrate habitat quantity and metapopulation dynamics should be treated as exploratory measures. The conclusions reached should therefore be viewed as

hypotheses, and any management recommendations based on these conclusions should be designed to test those hypotheses.
3. Where analyses are being conducted for multiple species, a subset of species should be selected for the intensive data collection needed to address the research objectives.

We note that item 1 might be a rare circumstance and not have widespread utility; most species will probably require specific study of their vital rates. Item 2 falls within the topic of adaptive management ("experimenting by doing") and can be applied to broad spatial extents using before/after experiments, where replication is non-existent or limited. For the guidance provided by Armstrong (2005) to be effective, however, we must add that investigators should take the trouble to identify, as thoroughly as possible, the actual structure of the population(s) of a species under study.

We realize that delineating population structure is not an easy task. Field biologists are aware of what it takes to design and implement a field study, especially one that involves broad spatial areas. Yet for most species, we do not have even a basic understanding of what their population structure might be. As such, a fundamental initial question in a study would be to determine if individuals of a particular species represent a disjunct distribution or a metapopulation with partially interbreeding subpopulations. That, in itself, is highly valuable information, providing far more insight into a species than a multitude of classic habitat studies. Once the species' fundamental distribution is understood, then a list of priority questions can be generated, according to the investigator's interest. These will probably be driven largely by the priorities of funding agencies: advancing theory, assessing rarity, and providing legal guidance (e.g., for endangered species). We discuss sampling design and methods in a forthcoming (2021) companion volume on applications. Below we offer guidance on ways to delineate biological populations.

A Broader Approach for Animal Ecology

Defining the biological population is complicated, because of discontinuities in distribution that are difficult to recognize. Often, little is known about individual animal movements across spatial and temporal scales. Regardless, we must first understand how the animals are assorted into meaningful, interacting groups of individuals before we can gather a proper sample. Population demography (e.g., age structure, survival) and viability have not been emphasized in most wildlife-habitat studies; abundance, density, or occupancy has generally been the focus. Measures of abundance or density have been emphasized, because they do not require us to know anything about the underlying population structure. The area needed to support viable demographic units, however, changes as we alter the size of our study areas. For example, calculating density does not mean that the derived value is of relevance to the animals being studied. In this section, we segue from the background and rationale developed above and provide guidance on approaches to studying animal ecology from a different perspective than has typically been followed in the past 50 or more years. Much of the following is summarized from Morrison and Block (2021), along with other literature describing methods for delineating biological populations.

The number of species across space is constantly changing, and it is simultaneously interacting with a varying amount of individuals within each of those species. As we revise spatial extents, these relationships likewise change. In addition, more-subtle interactions are occurring, such as changes the ratios for sex and, especially, age (Tri et al. 2012). Smallwood (2001) showed how enlarging the spatial extent of a sampling area increasingly captured functionally significant demographic units of different taxa, including both predators and prey (Fig. 5.5). Although this is a simple case, it clearly shows how the demographic relationships being expressed (e.g., density) change as the sampling area alters in size. Because

population viability considers how long a population will persist into a specified future time (see Chapter 1), we must know how a species is structured into populations if we are to appropriately design our studies. The primary reason we study animals and where they live (habitat) is to ensure their persistence into the future.

Morrison (2012:Fig. 1) expanded the example given by Van Horne (2002:Fig. 24.2; see also Morrison 2009:Fig. 3.5) that depicted how the abundance of a species could decline toward the edges of its range (Fig. 5.6). It is clear that where we sample along environmental gradients, relative to the geographic location of the species, will give us different relationships between a parameter and the response of that species (i.e., positive, negative, or neutral). We also see that the response of a species can change along a gradient even when distributions overlap.

Three major messages emerge from this. First, averaging across the gradient without regard to the underlying population structure (e.g., ecotypes) would result in a wide variance of the parameter estimates and give a false impression of how the species is responding to environmental conditions across space. Second, differences in how the species responds to environmental conditions is a critical consideration, especially if source-sink dynamics (Pulliam 1988) come into play. Here, a species may be able to reproduce in a particular area (i.e., "source," where reproduction exceeds mortality), and their offspring may disperse to another area where they are unable to breed (i.e., "sink," where morality exceeds reproduction). Averaging abundance or reproductive output across a broader landscape would provide misleading information on how a species is performing within a smaller area. A study occurring in just a source or a sink area would not represent the overall status of the population. This simple example shows, again, why determining the spatial structure of the species' population is a necessary initial step.

Third, we would not know how the individuals inhabiting our snapshot in space (study area) were

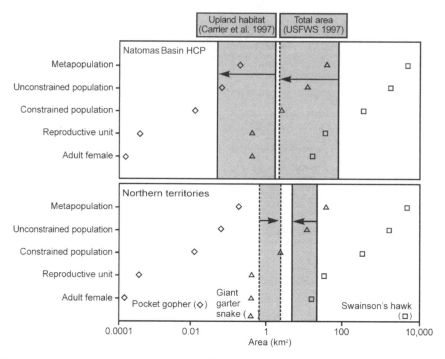

Figure 5.5   The areas needed to support functionally significant demographic units of animals vary with a project's spatial extent (size). Information on 3 species in the Natomas Basin Habitat Conservation Plan (*top*) and northern territories (*bottom*) was gathered for upland habitat (Carrier et al. 1997) and for the total area (USFWS 1997). The vertical lines correspond to habitat areas that are upland (*dashed lines*) or both upland and wetland (*solid lines*) at the start and end of the projects, as indicated by the respective origins and directions of the arrows. *Arrows pointing left* indicate a loss of habitat space, and *arrows pointing right* indicate gains. (Adapted from Smallwood (2001:Fig. 4), courtesy of K. S. Smallwood)

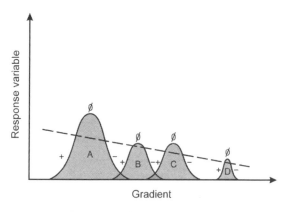

Figure 5.6   The response (e.g., density, productivity) of a species can vary across environmental gradients through space (*A–D*). Additionally, the portion of the gradient from which samples are taken can result in differing relationships (positive [+], negative [–], or neutral [Ø]), with the response being variable. The *dashed line* represents a hypothetical relationship when, for example, the underlying responses of different ecotypes are ignored.

being influenced by adjacent individuals, or how relationships would change over time as individuals move across our study area (a "neighborhood effect," per Dunning et al. 1992). By focusing on user-defined patches or communities within user-defined study areas, we are not, in any manner, accounting for these spatial and temporal dynamics. Thus our research mainly provides case studies that document—perhaps quite precisely—habitat relationships at a specific time and place. Nonetheless, because we do not capture the broader dynamics of our area of interest, we cannot accurately predict how habitat relationships will change, given future environmental conditions.

As has been developed in Chapter 1 and elsewhere in this book, many wildlife-habitat studies have focused on 1 or a few species found in a single location and measured a set of presumed habitat attributes that the researchers deemed important to the species under study (a shotgun approach, discussed above). Variables often relate to vegetation structure and composition, and perhaps topography, to describe a species' habitat (see Chapter 6). Data can be analyzed to simply describe the habitat, or to compare the portion that is used with the one that is available, in order to infer preference or selection. If done well, these studies might describe what a species is using at a particular place and time, but not why or how that picture will change across time under varying environmental conditions and as the dynamics of other species change. These studies would also fail to relate habitat to population parameters. Preference does not transparently relate to habitat quality, so we would learn little about the latter.

Such relatively vapid studies would be limited by the lack of a controlling logic that should determine both the scale of the research and the habitat analyses. As reviewed above, the process of identifying assembly rules is an attempt to describe broad patterns of species co-occurrence. We reiterate that we are not advocating any particular process in outlining how species might assemble in space and time. Rather, we are describing a structure that can be applied to better understand how and why species occur where they do. An assembly of different species, as a concept, provides a framework for a more holistic way to look at populations and their associated habitat use, because it forces an integration of what passes through various biotic and abiotic filters and constraints in studies conducted across space and over time and, thus, delves into niche factors, of which classic studies of habitat form a part. While a hierarchical view and a subsequent analysis of habitat selection describe a point in time, the framework of assembly rules starts by trying to understand how animals were able to settle within that habitat in the first place, given certain key filters and constraints. Characteristics of the selected environment—the habitat—can then be quantified across the various spatial and temporal scales relevant to the distribution of the focal species and the other species it interacts with in ways that modify its behavior, survival, and reproduction.

We are not suggesting sampling habitat across spatial scales per se. Rather, we are saying that environmental characteristics should be measured relative to the best-known information on a population's spatial structure. Descriptions of habitat will change as the location of the species (or group of species) varies, due to alterations in their biotic and abiotic environment (niche factors). This approach encourages the observer to consider a wider range of factors influencing behavior, including intra- and interspecific interactions, in a context broader than a user-defined (and convenient) study area, regardless of how spatially broad that area might be.

Above, we have discussed this approach in relation to golden-cheeked warblers. For this species, even using what most would consider a large study area (~40,000 km^2) would not provide any useful information on the population structure of these warblers (see Fig. 5.7). Because patch occupancy and abundance tend to increase based on geographic location within the range (Fig. 5.4), the location of an individual in such a sizeable study area would yield misleading results in relation to the conditions encountered by the species as a whole. We are not implying that focusing on a small study area does not yield valuable information on the ecology of an organism. In the case of golden-cheeked warblers, occupancy was shown to be related to vegetation patch size and the distance to adjacent patches (Fig. 5.8), although, as noted above, the magnitude of these relationships differed by the warblers' location within their range.

Populations and Their Delineation

In this book we have argued that the study of wildlife habitats is logically a part of the study of wildlife populations. There is convincing evidence both that studying an arbitrary subgroup from a population is unlikely to lead to patterns that can be extrapolated across the population as a whole, and that studies

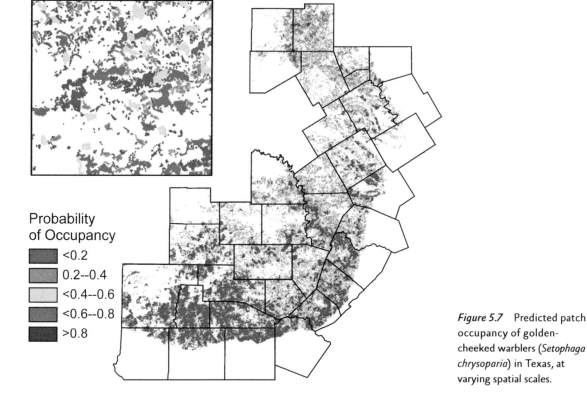

Probability of Occupancy

▨	<0.2
▨	0.2--0.4
▨	<0.4--0.6
▨	<0.6--0.8
▨	>0.8

Figure 5.7 Predicted patch occupancy of golden-cheeked warblers (*Setophaga chrysoparia*) in Texas, at varying spatial scales.

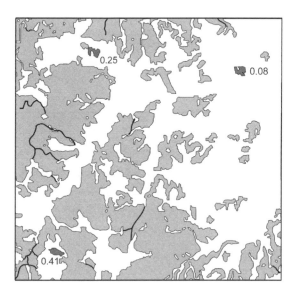

Figure 5.8 The probability of occupancy by golden-cheeked warblers (*Setophaga chrysoparia*) is related to the size of the area (i.e., patch) and its distance from other occupied areas. Areas shown in *dark grey* are examples of how occupancy probability decreases with increasing distance from other occupied (*light grey*) areas.

bridging multiple populations are likely to mask important local adaptations. In this chapter, we have expanded these ideas to the study of communities, arguing that such investigations should cover a spatiotemporal domain appropriate for a population-level study of all community members. Without this, communities become artificial, subjectively delineated constructs, wholly dependent on arbitrary spatiotemporal boundaries. While this position is, we feel, logically correct, robust approaches to delineate population are required to successfully implement these ideas.

The first step, naturally, is to coherently define a population. If we fail to do so, we cannot possibly hope to delineate it. One problem is that—similar to other widely used terms, such as community, ecosystem, landscape, and habitat—general definitions of population tend to be vague and unhelpful (as we explained in Chapter 1). For example, a visit to a biological dictionary (e.g., https://biologydictionary

.net/population/), will reveal the following: "A population is the number of organisms of the same species that live in a particular geographic area at the same time, with the capability of interbreeding." On the surface, this definition may seem reasonable.

If, however, we attempt to use this definition, or another of its many similar variants, to delineate populations in a landscape, its problems immediately become apparent. What, exactly, is a "particular geographic area"? What do we really mean by "at the same time"? And, perhaps most importantly, what does "capability of interbreeding" imply? For example, consider 2 groups of a migratory species that inhabit the same overwintering area but breed in separate locations. In the winter, they would seem to be a population: they are in the same place at the same time, and they are "capable" of interbreeding, though they're not going to do so. Most biologists would argue that, because the 2 groups do not interbreed, they do not constitute a single population. A good example of this thinking can be seen in the management of anadromous fish stocks, such as Pacific salmon (*Oncorhynchus* spp.). These fish mix in the pelagic environment for much of their lives, but they are largely genetically isolated because they breed in separate streams, returning primarily to the streams where they were born (Waples 1991).

In practice, there is a hierarchy of importance within the meaning of "population." Occupying the same areas at the same times does not define a population. On the other hand, interbreeding *does* define it, even if that population exists in separate spaces at all times and for all biological functions other than breeding. As such, we believe that populations are primarily a genetic construct. In this case, definitions associated with population genetics are probably more useful.

In genetics, populations are defined as breeding units. An idealized population is (1) closed (no migration), and (2) has random mating between all organisms within the closed population. For idealized models (e.g., random drift) to function correctly, additional requirements are assumed, such as non-

overlapping generations, sexual reproduction, and diploid organisms. Thus, in genetics, "population size" is the size of the reproducing population. Because ideal populations are vanishingly rare, the genetic population size for actual populations is thought of as being the effective population size, a concept based on a series of corrections that allow real (and, therefore, non-ideal) populations to be related to an ideal population, so expected behaviors can be modeled. Effective population size corrections almost exclusively decrease the overall population size.

Because effective population size is not an entity per se, but a set of correctional methods, it has 3 separate definitions, depending on the measured attribute of interest. "Inbreeding effective size" is based on the average inbreeding coefficient; "variance effective size" on the variance in the change of allele frequencies; and "eigenvalue effective size" on the loss of heterozygosity (Hartl and Clark 1989). Here we will more broadly (and less accurately) describe effective population size as the number of individuals you would need to have in an idealized population in order to exhibit the traits observed in a studied population. Effective population size cannot be directly measured. Rather, it is back calculated, based on observed population traits. For example, linkage disequilibrium, which is the non-random assortment of alleles within individuals in a population, occurs through random chance in small populations (the reasons for this are related to why you will seldom get 5 heads when flipping a coin 10 times). By assuming that all of the observed linkage disequilibrium is related to population size (a rather monumental assumption!), you can back calculate to the effective population size by measuring the linkage disequilibrium (Waples and Do 2008, 2010).

Genetics provides a solid definition of an idealized population and is at least a theoretical way to assess its size by using the idea of effective population size. For our purposes, however, delineating populations by means of this formal definition is not particularly useful. Problems emerge from the failure of real-world populations to exhibit the 2 primary constraints associated with an idealized population: closure and random mating. Outside of islands and isolated ponds, few systems are truly closed, in the sense that all breeding individuals are found exclusively inside the closed system. Usually systems are open to migrants. In many mammalian species, subadults—particularly subadult males—disperse widely, producing migrants that travel between areas. The ubiquity of this life history suggests that it conveys strong fitness advantages, probably by avoiding the negative genetic consequences of allele loss and inbreeding, which can affect small isolated populations. Classic genetic theory can only accommodate migrations if they are viewed as relatively rare events having no direct effect on the mating behavior of the population.

To understand the differences between real-world population processes and the ways in which these processes are modeled in population genetics, it is important to have a clear conception of the idealized models associated with population genetics. The form for an idealized population is commonly called an "urn model." In it, a jar (or urn) representing the spatiotemporal domain of a closed population is full of balls with different properties, representing individuals. Life-history processes—mating, reproduction, and death—occur be means of balls being randomly drawn from the urn. For example, 2 drawn balls could represent a successful reproductive event, in which case the properties of the selected balls (e.g., alleles) would be transferred to new balls (offspring), which would then be returned to the urn. Or a single ball could be removed from an urn and discarded, representing a death. Within the urn, balls are randomly intermixed and have no fixed spatial locations that might affect their being drawn.

It is important to note that virtually all of the material covered in a classic population genetics text represents properties of or variants on basic urn models. It is also vital to recognize that these models only formally simulate the breeding process. That is, it does not matter when or where organisms die;

what matters is the population composition when breeding occurs. This second axiom leads directly to a modification of the urn model, allowing the incorporation of migration across a metapopulation. In an idealized population model that includes migration, a metapopulation is visualized as a group of urns, each functioning independently, with the exception that occasionally a ball is removed from a particular urn and tossed into a different one (i.e., representing a migrant). If, however, a large number of migrants moved between the urns, this conceptualization breaks down. The populations are not fixed, independent entities. Rather, their behavior is strongly conditioned by the dynamics of the surrounding populations that share both alleles and organisms.

Nonetheless, some populations work in ways that allow this stylization to approximate actual population processes. A good example is found in the various species of Pacific salmon mentioned above. While their patterns in the ocean may be divergent, populations converge on spawning areas within their natal streams to breed. In these areas, breeding is fairly random. Further, the process is not perfect. Some salmon stray and end up in streams other than those where they were born. These strays then enter into the random breeding process, which, all and all, is very similar to the urn model that includes occasional migration. Note that the spatial and population dynamics outside the spawning area are not explicitly modeled. All that really matters is which individuals are produced on the spawning grounds and then make it back up the stream to successfully mate.

A critical factor in having a population that can be modeled using these classic assumptions is some type of mixing event that produces random mating and, therefore, random assortments of alleles within the population. In the case of salmon and many other fish species, this mixing takes place through their convergence on common spawning areas, but for several different species, it can occur at other points in their life history. For example, many marine organisms have a planktonic phase, where propagules are mixed by wind and currents. Although rock lobsters (*Panulirus cygnus*) have limited mobility as adults, they exist in extensive homogeneous populations, due to a planktonic larval phase (Thompson et al. 1996). While we make note of these examples, the ubiquity of urn models in population genetics is largely driven by their mathematical tractability, rather than their biological accuracy. Lacking a mechanism to produce a highly mixed population, the physical dimension of space is likely to impose constraints that preclude random mating. Because, in classic genetics, an ideal population is defined by random mating, non-random mating (due to constricted space) is destructive both to this conceptualization and to the idea of using it to delineate populations.

There is, however, a different and entirely divergent model for population genetics that was proposed in the 1940s (Wright 1943, 1946). In this model, mating and dispersal occur locally within a larger population. Unlike an urn, where a spatial location within a population is assumed to be irrelevant, here within-population breeding is decidedly non-random. This non-randomness is formally controlled by spatial location. Wright (1943, 1946) referred to the local zone in which breeding and movement activities occur as the "genetic neighborhood." Each neighborhood is centered on an individual and represents a probability distribution associated with mating and dispersal. Rather than having a pair of parents drawn randomly from the entire population, here it is assumed that the parents most likely came from a nearby space, with this likelihood decreasing with distance. A population is therefore made up of multiple individual-centered, overlapping neighborhoods, across which genes move diffusively. In an easier-to-contemplate finite approximation, this model is referred to as a "stepping-stone model" (Kimura and Weiss 1964; Weiss and Kimura 1965), where mating processes transfer genes to adjacent areas and move outward across the surface, affecting multiple generations.

While this model has generated much less attention, neighborhood types of mating processes are exceedingly common. In the world of plants, this fertilization structure (outside of deliberate or inadvertent manipulation) is inevitable and therefore ubiquitous. Consider a ponderosa pine (*Pinus ponderosa*) forest. Female gametes are in cones at the tops of trees, with male pollen cones at the base of the tree (an arrangement that limits self-pollination). Pollen is windborne—meaning that the density of pollen from any source will decrease diffusively with distance—and the resulting fertile seeds will simply cascade onto the forest floor in a cone-shaped frequency distribution immediately adjacent to the tree. Some seeds may be moved and cached by mammalian or avian species, but these activities will generally also be localized. (An exception are seeds distributed by human means.) Thus a sapling will usually be expected to have parents nearby. For plant species that use pollinator insects, such as bees, the localization of pollen will be limited by the working circles around the nests of potential pollinators. Other fertilization processes, such as seed-fall and nearby transport, are virtually identical.

This sort of genetic localization is common in mammalian species, as well. For example, the creation of territories produces a spatial structure in which mating is more likely with organisms occupying adjacent territories than it is with organisms from distant ones. Even for highly mobile modern humans, a person is more likely to marry a person from the neighborhood, town, state, nation, or continent where they were born than to marry an individual drawn from a world-wide population. These localization processes, whether they precisely follow Wright's (1843, 1946) model or not, lead to populations that are genetically divergent across space and form genetic gradients.

Ball et al. (2010) provided an example of how molecular genetics can help identify complexities in the relationships between populations at varying spatial scales. They employed individual-based clustering

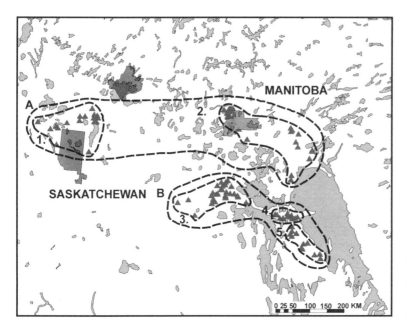

Figure 5.9   A consensus genetic population structure of caribou (*triangles*) showing coarse-scale (*lettered lines*) and fine-scale (*numbered lines*) genetic structuring. (Adapted from Ball et al. (2010:Fig. 4), courtesy of M. C. Ball)

(IBC) analysis, in which the operational animal units are individual genotypes rather than arbitrarily defined population units, thus providing an estimate of the number of populations and their spatial limits that was based on the genetic characteristics of each sample. Working with woodland caribou (*Rangifer tarandus caribou*) in central Canada and using several different analytical approaches, they showed how their strategy provided an effective means of delineating population structure and accurately assessing genetic diversity and connectivity (Fig. 5.9).

Returning to our problem of population delineation, it is clear that depicting a population that forms gradients will be different from one that encompasses largely discrete, homogeneous, randomly mating units. Thus the first step in thinking about how to outline a population for study is to think about what sort of population you are likely to be examining. Mistaking a gradient across a spatially extensive population for a series of discrete localized populations is common, precisely because, if you sample a gradient population intensively in relatively small separated areas, these exemplars can appear to be both divergent and locally homogeneous

(Schwartz and McKelvey 2009) and will have low calculated effective population sizes (Neel et al. 2013) Thus the attributes of your samples may be completely consistent with an isolated small-population model, even though these study areas exist within a fully continuous population. The conservation implications of these misunderstandings are large. A gradient population does not lose alleles more rapidly than a well-mixed population. (Actually, this happens a bit slower; see Maruyama 1972.) The apparently divergent characteristics and low sizes of measured effective populations that characterize local areas on a gradient surface do not indicate isolation and potential inbreeding, but, rather, the spatially correlated nature of gene flow. Alleles are derived from local sources in the short term, but in the long term, they are drawn from the population-level pool. Therefore, if you were to return to a site after 10 generations, the measured effective population size would be similarly small, but the genetic makeup entirely different. We provide additional details and examples of effective population size in Box 5.3.

For the purpose of population delineations, genetic gradients present real difficulties. Not only can gradients make it difficult to determine whether 2

Box 5.3. Calculations of Effective Population Size

Effective population size (N_e) is a multifaceted concept that refers to the number of breeding individuals in a population (N), corrected for a variety of conditions and factors that reduce the total size of the breeding population. The following are examples of such variants on the concept and their calculations (from Waples 2002).

Inbreeding-Structured Models of N_e

In these calculations, N is corrected for inbreeding rate (N_{ei}) or for changes in allele frequency (N_{ev}). The inbreeding effective population size, related to inbreeding rate, is

$$N_{ei} = \frac{(N*k)-2}{(k-1)+\frac{V_k}{k}}$$

and the variance effective population size, related to changes in allele frequency, is

$$N_{ev} = \frac{(N*k)-k}{1+\left(\frac{V_k}{k}\right)}$$

where N = population size in the parental generation, k = mean number of offspring produced per individual, and V_k = variance in k across individuals. The most conservative estimate for effective population size is the lower of the 2 calculated values (min [N_{ei}, N_{ev}]).

Spatially Structured Models of N_e

In these calculations, N is corrected for migration rates in subpopulation ($N_{et}W$), for intergenic genetic drift ($N_{et}N$), or for subpopulation extinction ($N_{et}N2$). The

effective population size corrected for sub-population migration rates (Whitlock and Barton 1997) is

$$N_{et}W = \frac{N_t}{((1+P)*(1-F_{st}))+\frac{N*P*F_{st}*n}{n-1}}$$

the effective population size corrected for intergenic genetic drift (Nunney 1999) is

$$N_{et}N = \frac{N_t}{((1+F_{is})*(1+F_{st})-(2*F_{is}*F_{st}))}$$

and the effective population size corrected for extinction of subpopulations is

$$N_{et}N2 = \frac{n}{4*(x+e)*F_{st}}$$

where N = population size in the parental generation, P = variance in productivity among subpopulations, F_{st} = Wright's (1943, 1946) measure of genetic differentiation among subpopulations, n = number of subpopulations, N_t = total number of individuals in the metapopulation, F_{is} = Wright's (1943, 1946) inbreeding statistic, x = migration rate among subpopulations, and e = rate of subpopulation extinction. Each is calculated, and then the lowest value of these 3 corrections (min [$N_{et}W$, $N_{et}N$, $N_{et}N2$]) is the best estimator of spatially structured effective population size.

Multiple-Factor Models of N_e

In this set of calculations (from Frankham et al. 2002), N is corrected for a variety of factors that would act to reduce the effective size of an idealized panmictic population. The factors include corrections due to sex ratio (N_eSex), variance in family size (N_eFam), variance in family size and mean family size (N_eFam2), fluctuations in population size over time (N_eFluc), and inbreeding (N_eInb).

The effective population size corrected for sex ratio is

$$N_e Sex = \frac{4 * N_{ef} * N_{em}}{N_{ef} + N_{em}}$$

the effective population size corrected for variance in family size is

$$N_e Fam = \frac{(4 * N) - 2}{V_k + 2}$$

the effective population size corrected for variance in family size and mean family size is

$$N_e Fam2 = \frac{(N * k) - 1}{k - 1 + \left(\dfrac{V_k}{k}\right)}$$

the effective population size corrected for fluctuations in population size over time is

$$N_e Fluc = \frac{t}{\Sigma\left(\dfrac{1}{N_{ei}}\right)}$$

and the effective population size corrected for inbreeding is

$$N_e Inb = \frac{N}{1 + F}$$

where N = total number of individuals, N_{ef} = effective number of breeding female parents, N_{em} = effective number of breeding male parents, V_k = variance in family size, k = mean family size, N_{ei} = effective population size in the ith generation, t = number of generations, F = inbreeding rate (ranging from 0 = none to 1 = completely inbred). As above, the best estimator is then taken as the lowest value from all of the corrections (min [N_eSex, N_eFam, N_eFam2, N_eFluc, N_eInb]).

Example

Using all of the above calculations, presume we have a population of N = 1000 breeding adults. For the inbreeding structured models of N_e, assume k = 1 and V_k = 2. The calculation results are N_{ei} = 499 and N_{ev} = 333, so the lower of the 2 values, 333, is the more conservative estimate of N_e for these factors.

For the spatially structured models of N_e, assume N_t = 100, P = 0.1, F_{st} = 0.0001, n = 2, F_{is} = 0.01, e = 0.1, and x = 10. The calculation results are $N_{et}W$ = 89, $N_{et}N$ = 99, and $N_{et}N2$ = 495, so the lowest of these values, 89, is the most conservative estimate of N_e for these factors.

Finally, for the multiple-factor models of N_e, assume N_{ef} = 25, N_{em} = 25, V_k = 2, k = 2, N_{ei} = [30, 20, 40, 30, 30 10], and F = 0.2. The calculation results are N_eSex = 50, N_eFam = 1000, N_eFam2 = 1000, N_eFluc = 22, and N_eInb = 833. The lowest of these values, 22, is the most conservative estimate of N_e for these factors.

Note the very wide range in the individual calculations and the correction factors considered in this example, each beginning with a total breeding population size of N = 1000. While each calculation is an authentic consideration of conditions that could influence the effective breeding size of the population, it is clear that none is the single, tangible, "real" population. Rather, each represents attention paid to different influences.

As to which calculation or set of calculations is the correct or best one to use depends entirely on first understanding the basic structure of the population in question, such as whether it constitutes a set of subpopulations, as well as having precise and unbiased estimates of each factor used in the calcula-tions. Further, how the population itself is delineated can affect the values used in the equations and, thus, their calculation outcomes. Clearly, usefully correcting for effective population size is more a matter of fundamentally understanding the species in question than it is of just producing calculations based on assumptions and best guesses of the parameters involved.

spatially disparate samples were drawn from the same population, but, in real-world landscapes, spatial gradients within habitats will be combined with habitat discontinuities, which can produce locally steeper gradients or actual population boundaries. Consider the earlier ponderosa pine (*Pinus ponderosa*) example. Ponderosa pines exist on drier sites, and in many areas of the western United States they are confined to lower elevations or occur on southerly aspects at higher elevations. As detailed above, breeding will be localized, producing gradients across the landscape, but these will occur at many scales. Within a south-facing slope, most alleles will probably be drawn from that slope, but some will originate from lower-elevation areas, fewer yet from the south-facing slope 1 drainage over, still less from a neighboring mountain range, and so forth. At some scale, these discontinuities will produce highly connected subpopulations; at a larger scale, a metapopulation; and at still larger scales, fully discrete populations.

Given that genetic gradients are expected across all scales except the largest, how might one analyze these patterns to delineate populations? One approach is to take samples from the genetic gradient and look for steep areas (subpopulations) and discontinuities (metapopulations or discrete populations). While methods to examine the slopes of genetic changes across surfaces have been developed (e.g., Guillot et al. 2005), they have not been commonly applied, because of the need for homogeneous sampling across large areas, and because the high variation that always exists between sampled individuals tends to obscure the larger features of the surface (Dupanloup et al. 2002). Clustering algorithms have been developed to allow population subdivision (e.g., Prichard et al. 2000), and they have been widely employed (at the time of this writing, Prichard et al. 2000 had been cited approximately 25,000 times). Using clustering algorithms, however, assumes a mating model that is close to the classic urn structure, with spatially discrete areas where mating is fairly random. Applying these tools naively to determine population structure across a population where mating is spatially localized, not surprisingly leads to erratic and questionable results (Schwartz and McKelvey 2009).

In the aggregate, these problems leave us with a bit of a conundrum. If genetic gradients are common, and their genetic effects are both hard to discriminate from a discrete subpopulation structure and difficult to interpret, how do we determine population ranges? And, if we cannot do this, how do we engage in population-based biology and management? It is clear that we cannot enter into the process of population delineation naively and expect post hoc statistical analyses of genetic patterns to reliably determine the extent of a population. Rather, we need to begin with the habitat and biology of the organism, as these factors are understood at the initiation of the study. Given the context of this book, the idea of delineating populations based on habitats initially appears to be paradoxical: we study populations to determine what constitutes habitat. Further, given that we have argued that populations are defined by breeding and, therefore, are primarily genetic entities, how can we ignore genetics in the process of population delineation?

We can, however, avoid these paradoxes by using the biology of the organism under investigation, separate from any specific habitat understandings, to delineate vegetation types that most likely have no useful role as habitat. That is, we may not know what habitat is, but we can reliably find areas that are not habitat (i.e., not used or selected by that organism). Cavity-nesting birds cannot exist in landscapes devoid of cavities and substrates to potentially provide cavities (e.g., grasslands). Species obligate to sagebrush (*Artemisia* spp.) cannot be found in areas lacking such shrubs. These sorts of basic understandings are known for most species of wildlife, although this knowledge may not be perfect. For example, an organism not observed exterior to dense mesic forests still might exist in other conditions.

Here, however, the genetic nature of population definition becomes a strength. In scaling our studies to populations, we are committing ourselves to the study of reasonably discreet breeding groups of individuals. Small populations that occasionally exist in unusual landscapes would only occasionally interact meaningfully with a study area scaled to a population. For this reason, a method to delineate local populations does not have to work globally. It only needs to reasonably set boundaries for an area of study, which is a much lower bar. Because of our focus on population and its genetic definition, we are interested in examinations of areas where organisms live and breed, not where they may occasionally occur. Thus small isolated fragments of vegetation that could, in principle, serve as habitat can be treated as

non-habitat in this context. Lastly, while genetic statistics are problematic for delineating population units, they are much more robust and useful for testing the validity of proposed population boundaries. Like most statistics, they do better when testing a priori hypotheses than serving as a post hoc "fishing trip."

We believe that this process of whittling down the landscape by removing areas of non-habitat will lead to biologically correct population and subpopulation delineations for many species (see Box 5.4). For unstudied species, this approach is somewhat problematic, as we arguably cannot reliably define what is non-habitat if we have little biological information on the relevant organism. Drawing conclusions from closely related species that are studied could also carry grievous errors. Even in these cases, however, we believe that a biologically based approach will provide the best guidance and offer superior results when compared with studies conducted at arbitrary spatiotemporal scales. In some cases, knowledge can be reasonably transferred from better-studied species that clearly exhibit similar traits and life histories, but caution is indicated whenever using them as similes for a poorly known species.

Methods exist and are being continually improved upon for identifying how animals are organized across space and over time. A first (albeit minimal) step in advancing our knowledge of animal ecology is acknowledging what we know and do not know about the organism(s) under study. Unfortunately, such a modest beginning is seldom applied in studies of animal ecology. Reviewing what other people have published is not equivalent to fully and critically evaluating what we know about a species' population structure in our proposed study. It is our hope that by making explicit statements regarding a biological population under consideration, this will force us to improve how we approach research. This should lead to more realistic and useful estimates of population parameters which, in turn, will have broader applicability.

Box 5.4. Population Delimitation for Spotted Owls (*Strix occidentalis*)

This organism (with 3 subspecies) exists throughout an area that extends from south-western Canada to northern Mexico. Across this extensive range there are considerable variations in the species' food habits and habitat associations. Nonetheless, there are common biological features that preclude the use of areas for spotted owl habitat. Spotted owls are perch-drop predators; to hunt they need trees and a reasonably open forest floor. Further, they do not construct nests and therefore need to find extant structures. In many forests they use secondary cavities. In some areas they use platforms associated with broken-topped trees; mistletoe brooms and similar branch structures; the nests of other birds; and, in the southwestern United States, occasionally ledges on cliffs. In many locales they need protection from the weather. In regions characterized by cold rain or snow, cavity nests become critical. In hot desert environments, the owls often perch under-neath heavy cavities and over water.

While these requirements would appear to encompass a broad variety of environments, there is an even wider assortment that cannot function as spotted owl habitat. These birds are unable to utilize grasslands and many types of shrublands; areas with small dense trees over extensive downfall (such as occur in stand-replacing fires roughly 20–40 years after the event); and, in colder environments, forests lacking sizeable cavities in large-diameter trees.

So, while identifying *all* habitats most likely would prove to be problematic, specifying extensive areas that cannot serve as habitat is relatively straightforward. Locales containing spotted owls that are isolated from other such places by non-habitat features potentially represent subpopulations. Additionally, we know certain things about the areal extents of habitat necessary for spotted owls to persist; home-range sizes less than approximately 250 ha have not been located (Zabel et al. 1992). Therefore, it is unlikely that spotted owls can live in isolated habitat areas smaller than this. These can also be viewed as non-habitat. Further, we know other aspects of spotted owl dispersal patterns: few dispersals greater than 80 km have been documented (Verner et al. 1992). Therefore, populations separated by non-habitat areas more than 80 km across probably can be considered to be separate populations.

LITERATURE CITED

Adams, D. C. 2007. Organization of *Plethodon* salamander communities: Guild-based community assembly. Ecology 88:1292–1299.

Agrawal1, A. A., D. D. Ackerly, F. Adler, A. E. Arnold, C. Cáceres, D. F. Doak, E. Post, P. J. Hudson, J. Maron, K. A. Mooney, M. Power, D. Schemske, J. Stachowicz, S. Strauss, M. G. Turner, and E. Werner. 2007. Filling key gaps in population and community ecology. Frontiers in Ecology and Environment 5:145–152.

Armstrong, D. P. 2005. Integrating the metapopulation and habitat paradigms for understanding broad-scale declines of species. Conservation Biology 19:1402–1410.

Ball, M. C., L. Finnegan, M. Manseau, and P. Wilson. 2010. Integrating multiple analytical approaches to spatially delineate and characterize genetic population structure: An application to boreal caribou (*Rangifer tarandus caribou*) in central Canada. Conservation Genetics 11:2131–2143.

Carrier, W. D., K. S. Smallwood, and M. L. Morrison. 1997. Natomas Basin Habitat Conservation Plan: Narrow channel marsh alternative wetland mitiga-tion. Report to Northern Territories, Inc., Sacramento, CA.

Clement, S., and R. J. Standish. 2018. Novel ecosystems: Governance and conservation in the age of the

Anthropocene. Journal of Environmental Management 208:36–45.

Cody, M. L., and J. M. Diamond, eds. 1975. Ecology and evolution of communities. Harvard University Press, Cambridge, MA.

Cohen, J. E. 1990. Food webs and community structure. Pp. 1–14 in J. E. Cohen, F. Briand, and C. M. Newman, eds. Community food webs: Data and theory. Springer-Verlag, New York.

Collier, B. A., J. E. Groce, M. L. Morrison, J. C. Newnam, A. J. Campomizzi, S. L. Farrell, H. A. Mathewson, R. T. Snelgrove, R. J. Carroll, and R. N. Wilkins. 2012. Predicting patch occupancy in fragmented landscapes at the rangewide scale for endangered species: An example of an American warbler. Diversity and Distributions 18:158–167.

Connor, E. F., and D. Simberloff. 1979. The assembly of species communities: Chance or competition? Ecology 60:1132–1140.

Diamond, J. M. 1975. Assembly of species communities. Pp. 342–444 in M. L. Cody and J. M. Diamond, eds. Ecology and evolution of communities. Harvard University Press, Cambridge, MA.

Dunning, J. B., B. J. Danielson, and H. R. Pulliam. 1992. Ecological processes that affect populations in complex landscapes. Oikos 65:169–75.

Dupanloup, I., S. Schneider, and L. Excoffier. 2002. A simulated annealing approach to define the genetic structure of populations. Molecular Ecology 11:2571–2581.

Esler, D. 2000. Applying metapopulation theory to conservation of migratory birds. Conservation Biology 14:366–372.

Fauth, J. E., J. Bernardo, M. Camara, W. J. Resetarits Jr., J. Van Buskirk, and S. A. McCollum. 1996. Simplifying the jargon of community ecology: A conceptual approach. American Naturalist 147:282–286.

Frankham, R., J. D. Ballou, and D. A. Briscoe. 2002. Introduction to conservation genetics. Cambridge University Press, New York.

Gotelli, N. J. 2000. Null model analysis of species co-occurrence patterns. Ecology 81:2606–2621.

Guillot, G., A. Estoup, F. Mortier, and J. F. Cosson. 2005. A spatial statistical model for landscape genetics. Genetics 170:1261–1280.

Hanski, I. 2004. Metapopulation theory, its use and misuse. Basic and Applied Ecology 5:225–229.

Hartl, D. L., and A. G. Clark. 1989. Principles of population genetics, 2nd edition. Sinauer Associates, Sunderland, MA.

Hein, A. M., and J. F. Gillooly. 2011. Predators, prey, and transient states in the assembly of spatially structured communities. Ecology 92:549–555.

Ings, T. C., J. M. Montoya, J. Bascompte, N. Blüthgen, L. Brown, C. F. Dormann, F. Edwards, D. Figueroa, U. Jacob, J. I. Jones, R. B. Lauridsen, M. E. Ledger, H. M. Lewis, J. M. Olesen, F. J. F. van Veen, P. H. Warren, and G. Woodward. 2009. Ecological networks—beyond food webs. Journal of Animal Ecology 78:253–269.

Kimura, M., and G. H. Weiss. 1964. The stepping stone model of population structure and the decrease of genetic correlation with distance. Genetics 49:561.

Kuhn, T. S. 1970. The structure of scientific revolutions, 2nd edition. University of Chicago Press, Chicago.

Leibold, M. A., M. Holyoak, N. Mouquet, P. Amarasekare, J. M. Chase, M. F. Hoopes, R. D. Holt, J. B. Shurin, R. Law, D. Tilman, M. Loreau, and A. Gonzalez. 2004. The metacommunity concept: A framework for multi-scale community ecology. Ecology Letters 7:601–613.

Leibold, M. A., and M. A. McPeek. 2006. Coexistence of the niche and neutral perspectives in community ecology. Ecology 87:1399–1410.

Levins, R. 1968. Evolution in changing environments: Some theoretical explorations. Princeton University Press, Princeton, NJ.

Looijen, R. C., and J. van Andel. 1999. Ecological communities: Conceptual problems and definitions. Perspectives in Plant Ecology, Evolution and Systematics 2:210–222.

Maruyama, T. 1972. Rate of decrease of genetic variability in a two-dimensional continuous population of finite size. Genetics 70:639–651.

Mathewson, H. A., J. E. Groce, T. M. McFarland, M. L. Morrison, J. C. Newnam, R. T. Snelgrove, B. A. Collier, and R. N. Wilkins. 2012. Estimating breeding season abundance of golden-cheeked warblers in Texas. Journal of Wildlife Management 76:1117–1128.

May, R. M. 1973. Stability and complexity in model systems. Princeton University Press, Princeton, NJ.

Miller, J. R. B., J. M. Ament, and O. J. Schmitz. 2014. Fear on the move: Predator hunting mode predicts variation in prey mortality and plasticity in prey spatial response. Journal of Animal Ecology 83:214–222.

Millstein, R. L. 2014. How the concept of "population" resolves concepts of "environment." Philosophy of Science 81:741–755.

Montoya, J. M., and R. V. Sole. 2003. Topological properties of food webs: From real data to community assembly models. Oikos 102:614–622.

Morrison, M. L. 2009. Restoring wildlife: Ecological concepts and practical applications. Island Press, Washington, DC.

Morrison, M. L. 2012. The habitat sampling and analysis paradigm has limited value in animal conservation: A prequel. Journal of Wildlife Management 76:438–450.

Morrison, M. L., and W. M. Block. 2021. Wildlife-landscape relationships: A foundation for managing habitats on landscapes. *In* W. F. Porter, C. J. Parent, R. A. Stewart, and D. M. Williams, eds. Wildlife management and landscapes: Principles and applications. The Wildlife Society, Wildlife Management and Conservation. Johns Hopkins University Press, Baltimore.

Morrison, M. L., B. A. Collier, H. A. Mathewson, J. E. Groce, and R. N. Wilkins. 2012. The prevailing paradigm as a hindrance to conservation. Wildlife Society Bulletin 36:408–414.

Morrison, M. L., and L. S. Hall. 2002. Standard terminology: Toward a common language to advance ecological understanding and application. Pp. 43–52 *in* J. M. Scott, P. J. Heglund, M. L. Morrison, J. B. Haufler, M. G. Raphael, W. A. Wall, and F. B. Samson, eds. Predicting species occurrences: Issues of scale and accuracy. Island Press, Washington, DC.

Myers, J. A., and K. E. Harms. 2009. Local immigration, competition from dominant guilds, and the ecological assembly of high-diversity pine savannas. Ecology 90:2745–2754.

Neel, M. C., K. McKelvey, N. Ryman, M. W. Lloyd, R. S. Bull, F. W. Allendorf, M. K. Schwartz, and R. S. Waples. 2013. Estimation of effective population size in continuously distributed populations: There goes the neighborhood. Heredity 111:189–199.

Nunney, L. 1999. The effective size of a hierarchically structured population. Evolution 53:1–10.

Ovaskainen, O., and I. Hanski. 2003. How much does an individual habitat fragment contribute to metapopulation dynamics and persistence? Theoretical Population Biology 64:481–495.

Pellet, J., E. Fleishmana, D. S. Dobkin, A. Ganderd, and D. D. Murphy. 2007. An empirical evaluation of the area and isolation paradigm of metapopulation dynamics. Biological Conservation 136:483–495.

Pritchard, J. K., M. Stephens, and P. Donnelly. 2000. Inference of population structure using multilocus genotype data. Genetics 155:945–959.

Pulliam, H. R. 1988. Sources, sinks, and population regulation. American Naturalist 132:652–661.

Rooney, N., and K. S. McCann. 2012. Integrating food web diversity, structure, and stability. Trends in Ecology & Evolution 27:40–46.

Root, R. B. 1967. The niche exploitation pattern of the blue-gray gnatcatcher. Ecological Monographs 37:317–350.

Schmitz, O. J., J. R. B. Miller, A. M. Trainor, and B. Abrahms. 2017. Toward a community ecology of landscapes: Predicting multiple predator-prey interactions across geographic space. Ecology 98:2281–2292.

Schwartz, M. K., and K. S. McKelvey. 2009. Why sampling scheme matters: The effect of sampling scheme on landscape genetic results. Conservation Genetics 10:441–452.

Simberloff, D., L. Stone, and T. Dayan. 1999. Ruling out a community assembly rule: The method of favored states. Pp. 58–74 *in* E. Weiher and P. Keddy, eds. Ecological assembly rules: Perspectives, advances, retreats. Cambridge University Press, Cambridge.

Smallwood, K. S. 2001. Linking habitat restoration to meaningful units of animal demography. Restoration Ecology 9:253–261.

Sokol, E. R., B. L. Brown, and J. E. Barrett. 2017. A simulation-based approach to understand how metacommunity characteristics influence emergent biodiversity patterns. Oikos 126:723–737.

Thompson, A. P., J. R. Hanley, and M. S. Johnson. 1996. Genetic structure of the western rock lobster, *Panulirus cygnus*, with the benefit of hindsight. Marine and Freshwater Research 47:889–896.

Tri, A. N., J. P. Sands, M. C. Buelow, D. Williford, E. M. Wehland, J. A. Larson, K. A. Brazil, J. B. Hardin, F. Hernandez, and L. A. Brennan. 2012. Impacts of weather on northern bobwhite sex ratios, body mass, and annual production in south Texas. Journal of Wildlife Management 77:579–586.

USFWS [US Fish and Wildlife Service]. 1997. Natomas Basin Habitat Conservation Plan. US Fish and Wildlife Service, Sacramento Field Office, Sacramento, CA.

Van Horne, B. 2002. Approaches to habitat modelling: The tensions between pattern and process and between specificity and generality. Pp. 63–72 *in* J. M. Scott, P. J. Heglund, M. L. Morrison, J. B. Haufler, M. G. Raphael, W. A. Wall, and F. B. Samson, eds. Predicting species occurrences: Issues of accuracy and scale. Island Press, Washington, DC.

Vellend, M. 2010. Conceptual synthesis in community ecology. Quarterly Review of Biology 85:183–206.

Verner, J., R. J. Gutiérrez, and G. I. Gould Jr. 1992. The California spotted owl: General biology and ecological relations. Pp. 55–78 *in* J. Verner, K. S. McKelvey, B. R. Noon, R. J. Gutiérrez, G. I. Gould Jr., and T. W. Beck, technical coordinators. The California spotted owl: A technical assessment of its current status. General Technical Report PSW-GTR-133. US Department of Agriculture, Forest Service, Pacific Southwest Research Station, Berkeley, CA.

Vögeli, M., D. Serrano, F. Pacios, and J. L. Tella. 2010. The relative importance of patch habitat quality and landscape attributes on a declining steppe-bird meta-population. Biological Conservation 143:1057–1067.

Waples, R. S. 1991. Pacific salmon, *Oncorhynchus* spp., and the definition of "species" under the Endangered Species Act. Marine Fisheries Review 53:11–22.

Waples, R. S. 2002. Definition and estimation of effective population size in the conservation of endangered species. Pp. 147–168 *in* S. R. Beissinger and D. R. McCullough, eds. Population viability analysis. University of Chicago Press, Chicago.

Waples, R. S., and C. Do. 2008. LDNE: A program for estimating effective population size from data on linkage disequilibrium. Molecular Ecology Resources 8:753–756.

Waples, R. S., and C. Do. 2010. Linkage disequilibrium estimates of contemporary N_e using highly variable genetic markers: A largely untapped resource for applied conservation and evolution. Evolutionary Applications 3:244–262.

Weiher, E., D. Freund, T. Bunton, A. Stefanski, T. Lee, and S. Bentivenga. 2011. Advances, challenges, and a developing synthesis of ecological community assembly theory. Philosophical Transactions of the Royal Society, B 366:2403–2413.

Weiss, G. H., and M. Kimura. 1965. A mathematical analysis of the stepping stone model of genetic correlation. Journal of Applied Probability 2(1):129–149.

Whitlock, M. C., and N. H. Barton. 1997. The effective size of a subdivided population. Genetics 146:427–441.

Wiens, J. A. 1981. Avian community ecology: An iconoclastic view. Pp. 355–403 *in* A. H. Brush and G. A. Clark Jr., eds. Perspectives in ornithology: Essays presented for the centennial of the American Ornithologists' Union. Cambridge University Press, Cambridge.

Wiens, J. A. 1989. The ecology of bird communities. Vol. 1, Foundations and patterns. Cambridge University Press, Cambridge.

Wiens, J. A. 2016. Ecological challenges and conservation conundrums: Essays and reflections for a changing world. John Wiley & Sons, West Sussex, UK.

Wright, S. 1943. Isolation by distance. Genetics 23:114–138.

Wright, S. 1946. Isolation by distance under diverse systems of mating. Genetics 31:39–59.

Zabel, C., G. N. Steger, K. S. McKelvey, G. P. Eberlein, B. R. Noon, and J. Verner. 1992. Home-range size and habitat-use patterns of California spotted owls in the Sierra Nevada. Pp. 149–163 *in* J. Verner, K. S. McKelvey, B. R. Noon, R. J. Gutiérrez, G. I. Gould Jr., and T. W. Beck, technical coordinators. The California spotted owl: A technical assessment of its current status. General Technical Report PSW-GTR-133. US Department of Agriculture, Forest Service, Pacific Southwest Research Station, Berkeley, CA.

6 — Managing Wild Animal Populations and Habitats in an Evolutionary and Ecosystem Context

> Collections of descriptive data are indispensable because they may lead to discovery of patterns (trends or recurrences). The patterns discovered, or the descriptive data per se, might lead to explanations (how) or hypotheses on cause (why).
>
> Guthery (2008:8)

Contemporary wildlife conservation is, more often than not, focused on priorities related to the here and now. We are faced with a litany of complex and often competing issues, such as declining populations and loss of diversity, along with the widespread fragmentation of habitats and populations, all in the context of an accelerating rate of extinctions caused by our prodigal societies and economies. Triage, whether we like it or not, is becoming more and more common as a means of establishing priorities for the management of wild animals and the habitats that sustain their populations (Arponen 2012), although the triage approach raises issues of unclear and competing conservation objectives (Vucetich et al. 2017).

While conservation crises have become all too common, it is important for animal ecologists to take a long and broad view of the forces that have led to where populations of wild animals are found today, as well as the places where they may be distributed in the future (Wiens 2016). The way to do this is to consider the distribution and abundance of animals from both evolutionary and ecosystem perspectives. These can then form the basis for developing management and conservation strategies that will provide the best chance of sustaining animal populations and habitats through space and time and, we hope, avoid the need for triage.

In this chapter, we summarize guidelines used for deciding what parts of a species' range might qualify for listing consideration under the US Endangered Species Act. We then present a set of suggestions for maintaining the long-term evolutionary potential of populations and, ultimately, of species lineages. Along the way, we correct a few misconceptions about metapopulations and species extinction and conclude with some challenges for applying a long-term evolutionary perspective to real-world, practical conservation action.

Of Ecotypes, Evolutionarily Significant Units, Distinct Population Segments, and Listing Decisions

One concept related to potential short-term adaptation is that of ecotype—a term that is often misused in the ecological literature or is employed with very different meanings, such as soil ecotype (sensu Baltensperger et al. 2017) or local landcover types and ecosystems (sensu Jorgenson et al. 2015). As we use it here, an ecotype is a portion or segment of a species, often a geographically distinct population, that is adapted to local environmental conditions differently than its overall parent species.

Ecotypes have been reported in populations of woodland caribou (*Rangifer tarandus caribou*) in the Selkirk Mountains of northern Idaho and southern British Columbia in Canada (Warren et al. 1996; Wittmer et al. 2005), garter snakes (*Thamnophis elegans*) in northern California (Palacios et al. 2011), killer whales (*Orcinus orca*) in the North Atlantic Ocean (Vongraven and Bisther 2014), as well as in other species. Understanding ecotypic variation can be vital for selecting the correct donor population for reintroducing a species, such as with bighorn sheep, where mismatches of ecotypes to translocation conditions can result in low recruitment (Wiedmann and Sargeant 2014).

A related concept, which has been broached in Chapter 1, is that of evolutionarily significant units (ESUs). Moritz (1994) suggested that ESUs should be defined as a geographically delineated set of individuals composing a unique monophyletic genome that is divergent from its parent population, so this delineation and conservation of ESUs can serve to maintain the natural genetic diversity of a species. Waples (1995) defined the ESUs of Pacific salmonids according to a population that was substantially reproductively isolated from other conspecific population units and represents an important component in the evolutionary legacy of the species. ESUs and similar designatable units have been used by the US Fish and Wildlife Service and the Committee on the Status of Endangered Wildlife in Canada for evaluating and listing species under the US Endangered Species Act and the Canadian Species at Risk Act (Green 2005; Haig et al. 2006).

Another concept used in defining populations for listing or recovery under the US Endangered Species Act is that of a distinct population segment, or DPS (DOI and DOC 1996). A DPS is based on the criterion that a population is discretely separate from other populations and thus has a demographic or genetic significance that differentiates it from other populations of the same species.

A third concept, also pertinent to US federal listings of species as threatened or endangered, is that of the significant portion of the range (SPR) of a species. The US Endangered Species Act, established in 1973, has defined an endangered species as one that is in danger of extinction throughout all or a significant portion of its range. But it took a special clarification from the Office of the Solicitor, US Department of the Interior, 33 years later to provide a specific definition of SPR, which formulated criteria for determining a specific population's contribution to the overall population by evaluating its degree of resiliency, redundancy, and representation.

Determining an appropriately specified unit—that is, which population—to address for potential listing, recovery, or other conservation actions, especially under the US Endangered Species Act, is no trivial matter. The determination procedure can follow 3 main steps, especially for addressing the legal criteria for DPSs and SPRs (Fig. 6.1). Step 1 (*A, top*) determines the range-wide status of and threats to a species, to ascertain if the listing consideration even needs to address a subset of the overall range. If it does, then step 2 (*B, middle*) is to determine the potential DPS status of that species by applying criteria to ascertain the degree of discreteness of a population segment, and, if that passes, to then judge the degree of significance of any discrete segment. Whether or not a species qualifies for DPS status, step 3 (*C, bottom*) can determine its SPR status by defining the portion of the species' range to be evaluated and then deciding on the significance of that portion by examining criteria for the species' resiliency, redundancy, and representation. Such a 3-step process has been used by US Fish and Wildlife Service for evaluating listing needs on a variety of species, and it could also be adopted by other conservation institutions.

Practical Considerations for Maintaining Evolutionary Potential

Beyond determining if a segment of a species' distribution deserves special consideration for conservation or a separate listing category, we can consider ways to bolster the evolutionary potential of that species. In this section, we offer practical considerations for managing conditions to maintain the evolutionary

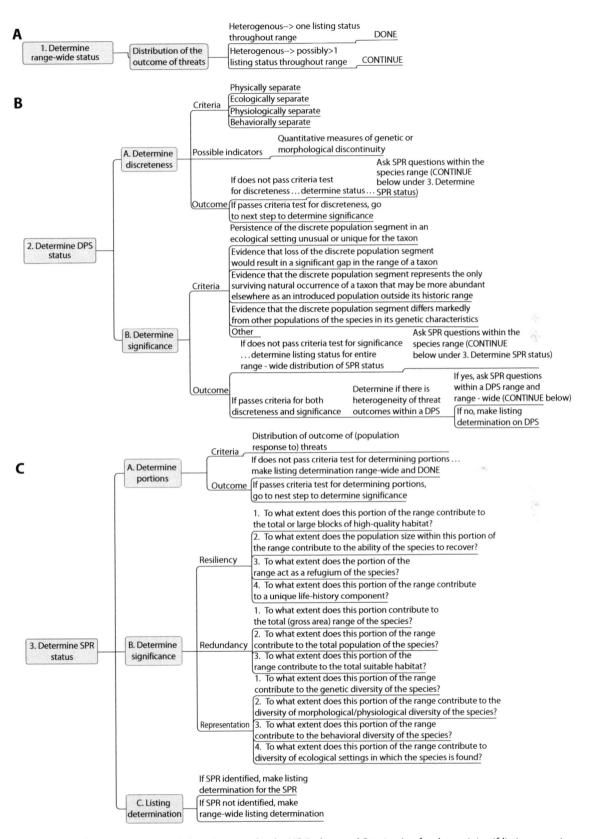

Figure 6.1 A decision tree summarizing criteria under the US Endangered Species Act for determining if listing a species should be considered (*A*) range-wide, (*B*) as a distinct population segment (DPS), or (*C*) as a significant portion of the range (SPR).

Table 6.1.   Guidelines for conditions to maintain evolutionary potential of species

Guideline for conditions to maintain	Purpose and intended effect
1. The full array of native environments, especially those in greatest decline	Provides for and restores environments and resources used and selected by native species, avoiding anthropogenic extirpations
2. The long-term viability of all species, with particular attention to at-risk species	Provides opportunities for populations to continue, well distributed, into future time periods
3. Environments and species' habitats along the peripheries of species' ranges	Allows conditions for potential character divergence, development of local adaptations, and parapatric subspeciation
4. Environments at the edge of a species' range of tolerance	Maintains the full diversity of individual variations for ecological flexibility within a population
5. Conditions for locally and regionally endemic species	Provides for continued existence of range-restricted and unique life forms
6. Locally endemic subspecies and locally adapted ecotypes	Maintains genetic diversity of gene pools within species
7. Cryptic or sibling species complexes	Provides for capacity of species assemblages to differentiate according to resource use and selection patterns
8. Conditions for stability and viability of disjunct populations	Provides opportunities for adaptations to local conditions
9. Population linkages where naturally occurring	Helps ensure gene flow and genetic variability
10. Polymorphic populations and other conditions of genetically diverse populations	Provides opportunities for within-population differentiation of forms with potentially varied adaptive responses to environmental changes

potential of wildlife entities. The prudent wildlife manager will recognize most or all of these situations in 1 form or another and be able to effectively merge them with other management needs.

Identifying the physical and environmental conditions needed to maintain the evolutionary potential of wild animals is a huge conservation challenge, primarily because there are so many moving parts and relationships in ecological and evolutionary systems (Wiens 2016). Much of it breaks down, however, to conservation or management actions that maintain or improve conditions sustaining taxonomic lineages. We seek to conserve and manage wild animals in the context of ecological domains that influence the trajectory of their populations, with an eye on maintaining their adaptive potential. Here, we suggest a set of 10 factors that can be used meet this goal (Table 6.1).

1. Maintain or restore the full array of native environments, especially those in greatest decline. This generally can help foster naturally heterogeneous environmental conditions and dynamics, within which the mechanisms of adaptation can proceed (Campos et al. 2010). An example is the prodigious work to save and restore the last remnants of tallgrass prairie in North America's Midwest.

Another example is in the inland west of the United States. Since the nineteenth century, 3 major vegetation conditions have experienced substantial declines in the interior Columbia River Basin: native shrublands have declined by 24 percent, old forests by 44 percent, and native grasslands by 63 percent (Marcot et al. 1998). Some species particularly associated with sagebrush steppes have also suffered losses, especially the iconic greater sage-grouse (*Centrocercus urophasianus*) (Davis et al. 2015).

2. Maintain the long-term viability of all species, with particular attention to at-risk species, which can include formally listed species and species of conservation concern in general. Viable populations, by definition, have sufficient distribution, abundance, and phenotypic and genotypic variations to allow them to adapt to changing conditions, within limits. Further, "all species" should include not just vertebrates, but also, as far as possible, fungi, lichens, bryophytes, vascular plants, and invertebrates, all of which collectively contribute to the diversity and productivity of healthy ecosystems. Species—including plants, invertebrates, and vertebrates that are naturally rare or become rare through human actions—can serve important functional roles (Walker et al. 1999; Miller et al. 2003).

Here we need to address the dark side of viability—extinction. Obviously, once a species is extinct, it has no chance for an evolutionary future. Or does it? There are 5 kinds of extinction, some of which certainly end the evolutionary lottery for a species, but others may just curtail some opportunities or, perhaps surprisingly, even hallmark their expression. The following is a summary of these 5 kinds of extinction, listed in no particular sequence.

The first kind is local extirpation, which is the local loss of members of a species, usually with the continued presence of that species elsewhere. Extirpations can occur for a wide variety of reasons, such as local overhunting, toxin(s) in the environment, effects of competition or predation from other species, pathogens or diseases, extreme weather events and environmental conditions, and so on. Local extirpations may eliminate some degree of the evolutionary potential of a species by reducing that species' geographic range and degree of genetic or behavioral variability within or among populations. Some species might bounce back from local extirpations, whereas for others it may be unclear what evolutionary potential has been lost, such as with the extirpation of American pikas (*Ochotona princeps*) from a number of sites in the Great Basin of the United States and in northeastern California (Beever et al. 2016). In a literature review, Maxwell et al. (2018) tracked 31 cases of local extirpations of plants and animals due to extreme weather events and found that interactions between such events, as well as other stressors, may have played a role in compromising the adaptive capacity of the relevant species.

The second kind considered here is global extinction, which is the loss of all members of a species, due to complete mortality. This situation eliminates any future evolutionary potential. Global extinctions can occur for several reasons, not least through human activity (Kerr and Currie 1995; Young et al. 2016). Some of the current concerns include extinc-

tions caused by anthropogenically accelerated climatic warming (Urban 2015).

The third kind, quasi-extinction, pertains more to modeling population viability. This term refers to the probability of a population dropping below a specified desired size (the quasi-extinction level), where it remains at non-zero but is still at high risk of full extinction. Rates of quasi-extinction have been modeled and calculated for populations of northern spotted owls (*Strix occidentalis caurina*) in the Pacific Northwest of the United States (Dunk et al. 2019), grey nurse sharks (*Carcharias taurus*) off the eastern coast of Australia (Otway et al. 2004), and many other species. When a population hits species-specific quasi-extinction levels, it can become far more susceptible to rapid declines. This is often referred to as an "extinction vortex," which is when a set of other factors combine to non-linearly increase the probability of extinction, such as from small-population random events (demographic stochasticity), unusual weather or other impacts on resources (environmental stochasticity), accelerated loss of genetic diversity (genetic drift), anthropogenic stressors, and other factors (Benson et al. 2019).

The fourth kind is functional extinction, which is when a population becomes so scarce that it no longer plays its function in its ecosystem (Sellman et al. 2016). Functional extinction of a single species or set of species can trigger cascading losses of other species. For example, if the functionally extinct species is a key pollinator or an obligate symbiont for another species, this might trigger what is known as "co-extinction" of the associated species. An example of this involves flying foxes (fruit bats of the family Pteropodidae), which are key dispersal agents of plant seeds in tropical forests. One study (McConkey and Drake 2006) showed that in areas where where flying foxes were scarcer, they dispersed less than 1 percent of the seeds they took, whereas elsewhere, where they were more abundant, they spread up to 58 percent of the seeds they consumed. Thus, in this case, flying foxes ceased to be functionally viable—that is, they became functionally extinct—as seed dispersers when their population densities were below a particular threshold, even though they were not necessarily locally extirpated. The conservation implication of this is that one might want to identify such thresholds, which may be greater than quasi-extinction ones, to help ensure that those key ecological functions are maintained. In another example, a revelatory twist arose in a study by Galetti et al. (2013), which suggested that the functional extinction of a set of seed-dispersing birds in the Brazilian Atlantic forest resulted in an unexpected evolutionary response: reductions in the seed size of a keystone species of palm tree that the birds had been dispersing. This change thus altered the evolutionary pathway and species composition of these tropical forests.

Lastly, the fifth kind of extinction is one seldom mentioned in the popular press or by those heralding an erstwhile mantra that species go extinct all the time, thereby suggesting that humans need not be concerned over their own adverse effects on biodiversity. This type is called "phyletic extinction," or "pseudoextinction," which occurs during evolution when a species becomes replaced by a different species form into which it has evolved, so the overall species lineage continues. Ironically, this kind of "extinction" is precisely what providing for long-term evolutionary potential is all about.

3. *Maintain environments and species' habitats along the peripheries of the species' ranges.* This tactic helps maintain unique environments with potentially different selective pressures than just what is within the core of the species' ranges. It is often at range peripheries where character divergence and parapatric subspeciation, or at least ecotypic differentiation, can occur (Sheth and Angert 2016). Peripheral conditions also allow floaters and sink habitats and can contribute to population stability (Lesica and Allendorf 1995; also see Furlow and Armijo-Prewitt 1995).

Another instance of local adaptation along a species' range periphery is in Mongolia, where peripheral populations of toad-headed agama lizards (*Phrynocephalus versicolor*) are found at the northern

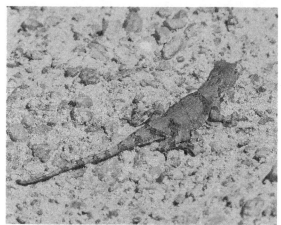

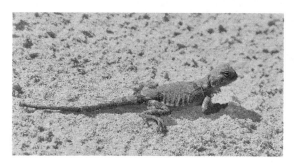

Figure 6.2 Morphs of the toad-headed agama (*Phrynocephalus versicolor*) in various locations of the Gobi Desert, Mongolia. Individual morphs are strikingly adapted to cryptically blend with their respective desert environments. (Photos by Bruce G. Marcot)

extend of their overall range in Asia. These lizards are sensitive to vegetation cover, topographic ruggedness, and wind speed (Yadamsuren et al. 2018). In Mongolia, the species has evolved a striking array of body patterns and coloration to blend well locally with their desert steppe environments (Fig. 6.2).

In a third example, Eckrich et al. (2018) found that conditions in the northern peripheral range of kit foxes (*Vulpes macrotis*) in southeastern Oregon did not provide sufficient ecological requirements, as well as core areas, for the species. They noted that work is needed to better determine the species' patterns of resource use and selection at their distributional periphery, in order to enhance conservation strategies. The literature provides many additional examples of unique adaptations, demographic responses, and conditions at the range peripheries of many species, including wild turkeys (*Meleagris gallopavo*) in southern Canada (Lavoie et al. 2017), sand lizards (*Lacerta agilis*) in central Europe (Henle et al. 2017), jaguars (*Panthera onca*) in Argentina (Cuyckens et al. 2017), fire salamanders (*Salamandra infraimmaculata*) in northern Israel (Blank et al. 2013), and American pikas (*Ochotona princeps*) in Idaho (Rodhouse et al. 2010).

In a study of the distributional patterns of wildlife in the interior Columbia River Basin, Marcot et al. (2003) calculated the density of the edges of species' ranges (as miles of edge per square mile of range) in each subbasin and overlaid their results by taxonomic group for all species. These findings, depicting hot spots in the density of range peripheries, surprisingly varied by taxonomic group (Fig. 6.3). Amphibian species' range peripheries were densest on the western edge of the study area, in the Columbia River Gorge, where warm, arid forests of ponderosa pine (*Pinus ponderosa*) and western larch (*Larix occidentalis*) on the east side interfaced with cool, moist forests of Douglas-fir (*Pseudotsuga menziesii*) and western hemlock (*Tsuga heterophylla*) on the west side. Other concentration areas included the east side of central Washington's Cascade Mountains, the arid pine-oak (*Pinus-Quercus*) forests of the Siskiyou Mountains in southern Oregon, the Snake River Canyon complex bordering Oregon and Idaho, and other spots.

In contrast, the highest concentration centers of range peripheries of mammal species occurred in other locations: along the eastern flanks of the Cascade Mountains in Washington and Oregon; farther south along the Snake River; and just east of Twin Falls, Idaho, where a number of species had either their northernmost distributions, such as pygmy rabbits (*Brachylagus idahoensis*), or their southernmost distributions, such as southern red-backed voles (*Clethrionomys gapperi*), demarcating a transition from southern to northern steppes in the region. Concentrations of range peripheries of amphibians and birds also differed, each with their own unique patterns. Steen and Barrett (2015) conducted a somewhat similar exercise, overlaying the distributions of amphibians and reptiles at state-wide scales, and concluded that peripheral-range occurrences of species are integral parts of their ecological communities and should have conservation value, as well as merit attention.

4. *Maintain environments at the edge of a species' range of tolerance, especially in the upper elevations, to allow for dispersal and range shifts during regional climate change.* This approach differs from the above focus on species' range peripheries and may be more difficult to implement, for the reason that less is known about a species' range of tolerance and selection of environmental conditions than about that species' general range distributions. One approach to identifying a species' range of tolerance is with climate niche modeling—a form of species distribution modeling—which determines and often maps climatic conditions, called the "bioclimatic envelope," that are correlated with species' distributions. The correlations are interpreted as conditions conducive to or selected and used by individuals, and they are employed to project a species' response to future conditions. Making provisions for the full range of a species' selected climatic and environmental conditions, particularly under changing climatic conditions, can be a helpful step to encourage that species' population persistence and its longer-term evolutionary potential.

Almpanidou et al. (2016) used climate niche modeling for loggerhead sea turtles (*Caretta caretta*) in the Mediterranean to determine that this species may be relatively secure under future scenarios that project general climatic stability in the region. Wüest et al. (2015) explored the dynamics of the range expansions of species under climatic differences

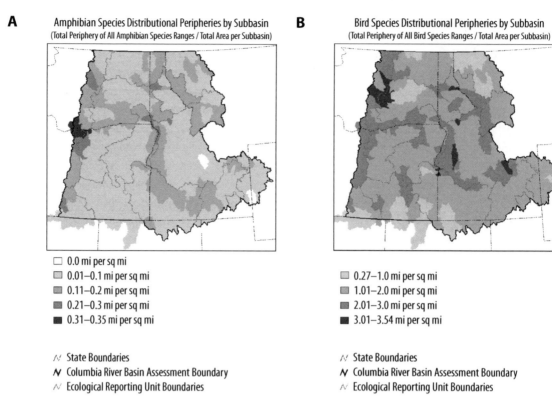

A Amphibian Species Distributional Peripheries by Subbasin
(Total Periphery of All Amphibian Species Ranges / Total Area per Subbasin)

☐ 0.0 mi per sq mi
▨ 0.01–0.1 mi per sq mi
▨ 0.11–0.2 mi per sq mi
▨ 0.21–0.3 mi per sq mi
■ 0.31–0.35 mi per sq mi

⋏ State Boundaries
Ν Columbia River Basin Assessment Boundary
⋏ Ecological Reporting Unit Boundaries

B Bird Species Distributional Peripheries by Subbasin
(Total Periphery of All Bird Species Ranges / Total Area per Subbasin)

▨ 0.27–1.0 mi per sq mi
▨ 1.01–2.0 mi per sq mi
▨ 2.01–3.0 mi per sq mi
■ 3.01–3.54 mi per sq mi

⋏ State Boundaries
Ν Columbia River Basin Assessment Boundary
⋏ Ecological Reporting Unit Boundaries

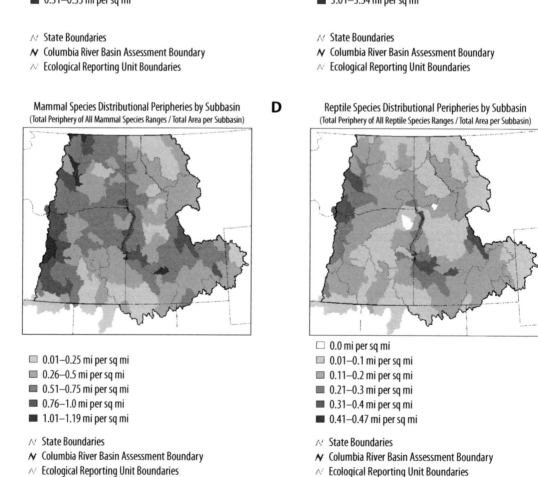

C Mammal Species Distributional Peripheries by Subbasin
(Total Periphery of All Mammal Species Ranges / Total Area per Subbasin)

▨ 0.01–0.25 mi per sq mi
▨ 0.26–0.5 mi per sq mi
▨ 0.51–0.75 mi per sq mi
▨ 0.76–1.0 mi per sq mi
■ 1.01–1.19 mi per sq mi

⋏ State Boundaries
Ν Columbia River Basin Assessment Boundary
⋏ Ecological Reporting Unit Boundaries

D Reptile Species Distributional Peripheries by Subbasin
(Total Periphery of All Reptile Species Ranges / Total Area per Subbasin)

☐ 0.0 mi per sq mi
▨ 0.01–0.1 mi per sq mi
▨ 0.11–0.2 mi per sq mi
▨ 0.21–0.3 mi per sq mi
▨ 0.31–0.4 mi per sq mi
■ 0.41–0.47 mi per sq mi

⋏ State Boundaries
Ν Columbia River Basin Assessment Boundary
⋏ Ecological Reporting Unit Boundaries

Figure 6.3 Concentrations of range distributional peripheries of amphibians (*A*), birds (*B*), mammals (*C*), and reptiles (*D*) of the interior Columbia River Basin, by subbasin. (Reproduced from Marcot et al. (2003), courtesy of the US Department of Agriculture, Forest Service)

among continents. They identified how biotic interactions—as well as genetic, developmental, and functional constraints—can guide niche diversification. Copeland et al. (2010) modeled the bioclimatic envelope of wolverines (*Gulo gulo*) across their distribution in western North America and concluded that future reductions in spring snow cover, under a warming climate, would probably shrink habitat for that species. Bell and Schlaepfer (2016) cautioned that future projections of species distribution models, particularly extrapolations to new or novel climatic conditions, may be fraught with error unless the models can be independently evaluated, using training data.

Another approach to determining a species' range of tolerance of environmental conditions is through bioenergetics modeling. For instance, Bartelt et al. (2010) modeled the environmental biophysics of western toads (*Anaxyrus* [= *Bufo*] *boreas*) in southeastern Idaho. They determined that regional climate warming will induce increasingly less-suitable temperature and moisture conditions, limiting the toads' movement and activity patterns among their habitats, thus potentially curtailing their capacity for adapting to climate change.

5. *Maintain conditions for locally and regionally endemic species.* Perhaps more obvious than not, conserving conditions and populations of endemic species can provide opportunities for local adaptation and evolutionary potential. These are the unique signatures of local ecosystems and, once lost, are gone forever. Together with at-risk (threatened) species, targeting endemic species for conservation can help maintain a significant proportion of biodiversity in an area (Ceballos et al. 1998; Posadas et al. 2001; Bonn et al. 2002).

There can be several causes for why some species are highly range-restricted and locally endemic. Endemism hotspots can demarcate areas of stable climatic refugia (Harrison and Noss 2017). Some paleoendemics have persisted in refugia locations with suitable conditions, although much of their previous distribution area has contracted or vanished. Some neoendemics may represent a recent divergence of species' lineages in situations with isolated conditions, range-periphery

character divergences, or other contexts. Ibarguchi (2014) noted that arid environments could contain unique endemic species and "evolutionary novelty," such as in the cold, arid lands of the Southern Cone of South America, and deserved conservation priority.

Islands often harbor endemics and present challenges for their conservation under conditions with anthropogenic stressors of habitat loss, as well as, in oceanic islands, rising sea levels from climate change (Wetzel et al. 2013). In continental locations, some isolated environments are considered to be island conditions. An example is the "sky island" mountaintops of the Western Ghats range in southern India, where isolated endemic populations of threatened bird species, such as white-bellied shortwings (*Myiomela albiventris*), occur (Purushotham and Robin 2016). These researchers analyzed the song characteristics of the shortwings and determined that the species' population isolates are the result of both ancient and recent separations, the latter from habitat fragmentation caused by deforestation.

6. *Maintain locally endemic subspecies and locally adapted ecotypes.* Another key to enhancing the evolutionary potential of species is to ensure that within-species variations are maintained. These variations include subspecies, ecotypes, evolutionarily significant units (see Chapter 1), and others (described below), since they constitute unique life forms adapted to local conditions and contribute to the overall genetic diversity of a species' gene pools.

Two examples of locally adapted and highly range-restricted subspecies from the inland west of the United States are potholes meadow voles (*Microtus pennsylvanicus kincaidi*) and White Salmon pocket gophers (*Thomomys talpoides limosus*). Both of these are also instances of subspecies occurring at the periphery of their parent species' ranges. Greenberg and Maldonado (2006) noted that the West Coast of the United States contains an anomalously higher number of endemic vertebrates, including subspecies such as salt marsh harvest mice (*Reithrodontomys raviventris*), than do other global mid- to high-latitude coasts. Norway's Svalbard archipelago in the High

Arctic contains several endemic subspecies of conservation interest, including Svalbard rock ptarmigans (*Lagopus muta hyperborea*) (Pedersen et al. 2014) and Svalbard reindeer (*Rangifer tarandus platyrhynchus*) (Zhao et al. 2019). The Hawaiian Islands and the islands of New Zealand are well known for their wide arrays of endemic subspecies and species.

7. *Maintain cryptic or sibling species complexes.* Cryptic or sibling species, sometimes termed "sister species," are sets of species that have evolved and generically diverged but still appear to be identical or nearly identical, although they may still be undergoing further divergence in their phenotypes, genetics, and behavior. Some cryptic species complexes have only recently become differentiated, whereas others may be known from the paleorecord. Such complexes contribute to the genetic diversity that underlies an adaptation to changing or diverse environmental conditions. In some cases, a sibling species may be endangered and require listing and recovery efforts (e.g., McElroy et al. 1997).

Examples of extant cryptic species complexes include two North American flying squirrel species (*Glaucomys volans* and *G. sabrinus*), along with a recently discovered third species (*G. oregonensis*), for which Arbogast et al. (2017) proposed the common name Humboldt's flying squirrel. Many other examples abound in the literature, including cryptic species complexes of crossbills (*Loxia* spp.), forest flycatchers, forest warblers, and more. Alström et al. (2016) discovered that South Asia's plain-backed thrushes (*Zoothera mollissimia*) actually constitute a cryptic complex of 2 species.

Again, island systems are sometimes hotbeds of adaptive radiation for cryptic species complexes (Pontarp et al. 2015). Striking examples of this include the complex of 7 species of lava lizards (*Microlophus* spp.) of the family Tropiduridae in the Galápagos Islands of Ecuador (Fig. 6.4), and the complex of no less than 34 species of *Litoria* tree frogs in Australia (Robinson 1999). Stroud and Losos (2016) noted that sometimes evolutionary change, including instances of adaptive radiation, can proceed rapidly, along short time scales of ecological processes, leading to evolutionary diversification.

8. *Maintain conditions for the stability and viability of disjunct populations.* Disjunct populations are geographically and, potentially, genetically isolated from other locations in their species' range. Disjunct populations may have developed into locally adapted ecotypes or subspecies, or they may at least harbor the evolutionary potential for doing so.

Populations may become disjunct naturally, as happens in isolated refugia, or through human fragmentation of their habitats. In the former condition, artificially connecting populations that are otherwise naturally isolated, or inappropriately translocating individuals among or into disjunct populations, may be detrimental (Storfer 1999; Russell et al. 2002) and disrupt their evolutionary potential by genetically swamping the gene pool. The argument is the same for maintaining separation of small isolated patches of vegetation, wetlands, or other environments that can serve as critical refuges and stepping-stones for many organisms (e.g., Saunders et al. 1987; Semlitsch and Bodie 1998). Such reconnections or translocations may be beneficial to populations, however, if human activities caused the disjunction.

In some cases, disjunct populations are modern day artifacts of once-contiguous species distributions now separated by changing climatic conditions and shifts in vegetation and ecosystem conditions. India has a number of such examples. There, some wildlife species are found in disjunct locations in the country's northeastern states and in the Western Ghats mountains of the south, including little spiderhunters (*Arachnothera longirostra*), Asian fairy bluebirds (*Irena puella*), and mountain hawk eagles (*Spizaetus nipalensis*). Some of these species may also be examples of secondary dispersal and colonization, although only a fossil or subfossil record would tell us for sure.

In the example of species' patterns in the interior Columbia River Basin, Marcot et al. (2003) mapped the density of various wildlife taxa with disjunct populations (Fig. 6.5). A hot spot containing the greatest number bird species with disjunct populations oc-

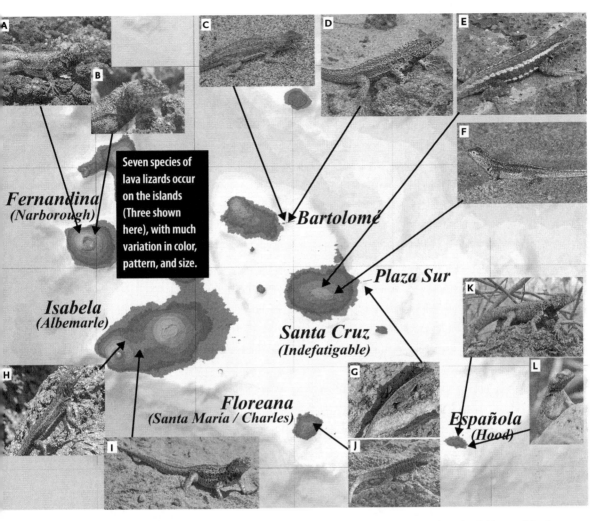

Seven species of lava lizards occur on the islands (Three shown here), with much variation in color, pattern, and size.

Figure 6.4   Examples of the adaptive radiation forms of lava lizard species (*Microlophus* spp.) in the Galápagos Islands, Ecuador. Seven lava lizard species occur on the islands, with much variation in color, pattern, and size. Three of the species are shown here: Galapagos Lava Lizard (*M. albemarlensis*), A–I; Floreana Lava Lizard (*M. grayi*), J; Espanola Lava Lizard (*M. delanonsis*), K–L. (Photos by Bruce G. Marcot. Map of the Galápagos Islands via Wikimedia Commons. Creative Commons Attribution–ShareAlike 3.0 Unported License.)

curred in the Steens Mountains of southeastern Oregon (*C*). For mammals, it was farther north, in the Great Basin steppe (*D*); for reptiles, along the Columbia River Gorge (*B*). Amphibians had no specific concentration centers throughout the region (*A*). Among the reptile species, ringneck snakes (*Diadophis punctatus*) had a disjunct distribution (Fig 6.6), with some populations appearing to be relatively small and isolated (also see the next topic, below).

Whether such mapped distributions reflect actual disjunct patterns, or are artifacts of incomplete field surveys, remains to be determined for a number of furtive species. Also of interest is discovering if environmental characteristics, including climatic conditions, vary significantly among locations with apparently disjunct populations. If they do, it might suggest either that the overall species is flexible enough to deal with a range of conditions, or that the disjunct populations

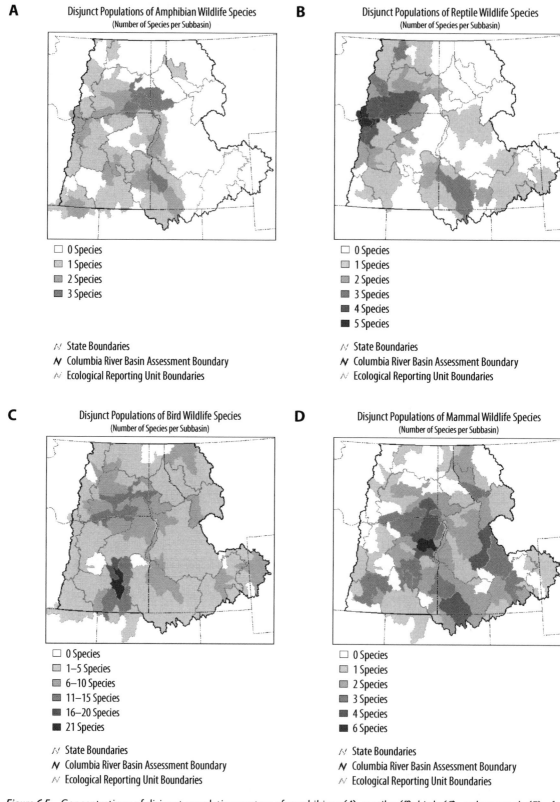

Figure 6.5 Concentrations of disjunct population centers of amphibians (*A*), reptiles (*B*), birds (*C*), and mammals (*D*) of the interior Columbia River Basin, by subbasin. (Reproduced from Marcot et al. (2003), courtesy of the US Department of Agriculture, Forest Service)

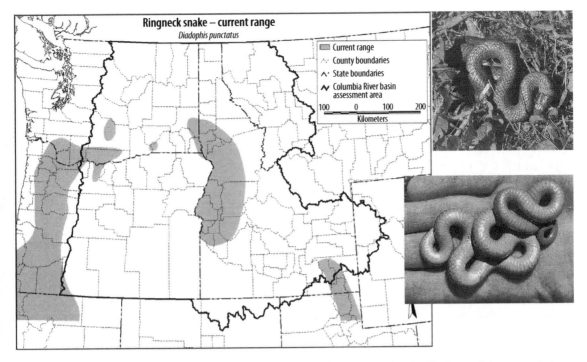

Figure 6.6 Ringneck snakes (*Diadophis punctatus*) are an example of a species with a distribution as disjunct populations, here shown for the interior Columbia River Basin. (Map reproduced from Marcot et al. (2003), courtesy of the US Department of Agriculture, Forest Service; photos by Bruce G. Marcot)

are locally adapted to the different conditions. Hallfors et al. (2016) explored this situation by creating MaxEnt-based species distribution models of 2 disjunct populations of Karner blue butterflies (*Lycaeides melissa samuelis*) in the Great Lakes region of the United States. They found that climatic characteristics varied significantly between the populations and concluded that these populations were locally adapted to the different conditions, although these researchers did not perform translocation experiments by swapping individuals between the sites to determine the resulting fecundity and survival of the populations. The authors cautioned that population-specific models could reveal more about local adaptations and conditions than whole-species models.

In a similar study, but with a very different outcome, Shipley et al. (2013) developed MaxEnt niche models of 2 disjunct populations of painted buntings (*Passerina ciris*) in the southern United States, separated by a 500 km gap. They determined that the models did not differ, suggesting that the gap was not bioclimatic in origin. It may be likely that the Gulf of Mexico currently serves as a dispersal barrier between the populations. Again, the lesson here is not to jump to conclusions about local adaptations in apparently disjunct populations.

9. *Maintain population linkages where they are naturally occurring.* This guideline is essentially the inverse of sustaining naturally occurring, locally adapted disjunct populations or ecotypes. Preserving such linkages among populations aids in retaining the species' overall genetic and demographic viability, including helping to ensure a sufficient effective population size for species persistence. Linkages assist in reducing risks of local extirpation from environmental disturbances, demographic stochasticity, inbreeding depression, or other threats. Habitat corridors and other linkages among populations can take many forms, at many scales, depending on the species' requirements and local conditions (Harris 1988; Hudson 1991).

Again, it may not be wise to link naturally occurring disjunct and isolated populations (see above), as this could result in genetic swamping, a loss of genetic identity for individual populations, and a loss of potential for adapting to different conditions.

Some researchers have employed dispersal modeling to determine linkages among populations. Fattebert et al. (2015) used a second-order resource-selection function model of structural connectivity of habitats, based on GPS data, to demonstrate that leopards (*Panthera pardus*) could use dispersal habitat peninsulas in a human-dominated landscape in South Africa. They concluded that their modeling approach would be useful in planning landscape conservation for wide-ranging carnivores. As we have pointed out in Chapter 1, however, interactions of individuals between populations or subpopulations do not necessarily mean that a metapopulation structure exists. The concept of metapopulation structure may have been overused in the literature, with little empirical testing done.

10. *Maintain polymorphic populations and other conditions of genetically diverse populations.* This final guideline for conditions to help preserve the evolutionary potential of species pertains to identifying and maintaining environmental conditions for polymorphic populations (those that have more than a single phenotype in the same population). In a way, polymorphic populations—which express diversity within a population (Wiens 1999)—are the inverse of cryptic or sibling species complexes, which are identical or near-identical phenotypes among a set of species. In some circumstances, polymorphic populations can form the basis for adaptive radiation and the evolution of subspecies and species complexes (Smith and Skulason 1996) and are what Butler and Mayden (2003) termed "cryptic biodiversity." Such genetically diverse populations have the potential for evolving adaptive behaviors and speciation (Mallet and Joron 1999). The loss of genetic diversity from a population may decrease the fitness of a species (Avise 1995), as was documented by Hitchings and Beebee (1998) in common toads (*Bufo bufo*).

One example of a polymorphic population is that of Stewart Island (Foveaux) shags (*Leucocarbo chalconotus*, syn. *L. stewarti*), a locally endemic species of cormorants found in New Zealand. In the same breeding population, individuals occur as 2 morphs: a completely dark bronze morph, and a black-and-white pied morph (Fig. 6.7). The 2 morphs are indicative of a balanced polymorphism, with neither becoming the sole expressed form. It is unclear how this evolved; what advantage, if any, either or both morphs enjoy; and what may become of this situation in the future, especially under conditions of environmental change. It seems prudent, however,

Figure 6.7 Stewart Island (Foveaux) shags (*Leucocarbo chalconotus*, syn. *L. stewarti*) of New Zealand are an example of a population with balanced polymorphism. Dark (bronze) and black-and-white (pied) morphs both occur in the same breeding population. (Photo by Bruce G. Marcot)

to maintain such conditions for the population as a reservoir of adaptive and evolutionary potential.

Anthony et al. (2008) studied polymorphisms in a population of red-backed salamanders (*Plethodon cinereus*), where some individuals are striped and others are not. Under experimental conditions, they found that the diets of striped individuals were significantly more diverse and included the selection of more useful prey than the diets of unstriped individuals. On the other hand, the latter individuals were better adapted to warmer environments. The 2 forms apparently have developed some degree of ecological separation, but they interbreed frequently enough to offset further divergence, at least under current environmental conditions. Again, the point here is maintaining the potential for evolutionary adaptation, which may not manifest itself until environmental conditions change.

Although the above list focuses, in part, on maintaining naturally segregated conditions (range peripheries, disjunct populations, etc.), there are many cases where the evolution of new species can potentially occur without geographic barriers that isolate populations and prevent gene exchanges. Such situations may be indicated by the occurrence of ecotypes, sympatric polymorphisms, and sympatric sibling or cryptic species. A number of selection forces can serve to differentiate species sympatrically, such as bacterial agents, mimicry, changes in a single pigment gene or the song of a bird, or rapid evolution of a sperm-surface protein (Morell 1996). Other conditions worthy of attention for maintaining long-term evolutionary potential of species' lineages include understanding the roles of hybridization, diseases and parasitism; effects of commensalisms with and dispersal by humans; adverse effects of exotic species; and other topics we have covered in this book. Many of these factors are left to species-specific evaluations.

The management of species, or of segments of a species' distribution, for evolutionary potential is not a new concept, as noted in the earlier reference to Frankel and Soulé (1981). For example, it has been used under the US Endangered Species Act to delineate and manage threatened salmonid stocks. For this purpose, the formal concept of an evolutionarily significant unit has been defined as a population that (1) is substantially reproductively isolated from other conspecific population units, and (2) represents an important component in the evolutionary legacy of the species (Waples 1995; see also Pennock and Dimmick 1997; DeWeerdt 2002).

Karl and Bowen (1999) suggested using the term "geopolitical species." Through mitochondrial DNA research, they discovered that endangered black sea turtles (*Chelonia agassizii*) were not distinct from green sea turtles (*C. mydas*), yet the designation of and conservation focus on the former has persisted for over a century, based on geographic and political considerations. They concluded that black sea turtles should continue to be the focus of endangered species conservation, because of their morphological diversity and the possibility of them being an incipient species with novel adaptations. This corresponds well with our call for appreciating and conserving novel or endemic subspecies, morphs, and other genetically unique portions of a species' lineage.

Can Ecosystems Be Managed?

Our relatively new era of ecosystem management (e.g., Zorn et al. 2001; Robbins 2012) presumes many things (Butler and Koontz 2005), such as that we know enough about how ecosystems work—including the roles and interactions of wildlife—to effectively depict, predict, and control the effects of our activities. The Wildlife Society recommended a set of performance measures for tracking the effects of human activity on ecological sustainability (Haufler et al. 2002). Despite considerable study over a great deal of time, however, much of the science underlying our ability to predict ecosystem responses to change is wanting (Real and Brown 1991). At best, we can be sure that many of our activities will indeed affect ecosystems, but determining to what end is often a best guess or is pertinent to only a small portion of what constitutes an ecosystem, such as the abundance of well-studied vertebrate populations or the productivity of rangelands over time.

Conservationists have struggled to agree on an unambiguous definition of ecosystem management (Grumbine 1994; Christensen et al. 1996; Cortner et al. 1999; Robbins 2012). Despite this struggle, some clarity in the concept of ecosystem management can be gained by separating science from goals and policy. Goals and policies are political in nature, and even though they can be informed by science, often they are not. From the scientific perspective, asking the question, "What is an ecosystem and how can it be mapped for management purposes?" is a useful start for implementing the philosophy of ecosystem management in a particular context. Much of the controversy surrounding ecosystems as a unit of study and management is really about identifying an ecosystem as a definable unit. Here we will briefly review this discussion, because it is important to understand the ongoing controversy prior to using the term "ecosystem," and especially before designing an ecosystem study and guidelines for ecosystem management.

Defining and choosing indicators for ecosystems would be possible if the science of ecology was able to provide simple, rigorous models for describing and predicting the status and dynamics of ecosystems. Our knowledge, however, is largely inadequate to identify the most important variables (Keddy et al. 1993). Likewise, Tracy and Brussard (1994) argued that we do not yet know which ecosystem processes will be the most important to study, index, and preserve. They noted that ecosystems have been arbitrarily defined study units that ranged from water droplets to the entire biosphere! As summarized by Peters (1991:91), the term ecosystem is a "multidimensional, unlimited, relativistic" entity representing the environment.

Recent efforts have helped bring clarity and definition to some ecosystem-related concepts, such as biotic integrity (Wellemeyer et al. 2018), ecosystem collapse (Bland et al. 2018), ecosystem resilience (Sasaki et al. 2015), and others. Still, despite various approaches to delineating and mapping ecoregions as proxies for ecosystems (e.g., Sarr et al. 2015), ecologists have not yet adopted a single, general classification system for (non-systematic) ecological units such as species assemblages, ecological communities, and ecosystems—except, perhaps, for biome, which is too large and crude to be a useful scale of study and management (Orians 1993). In ecology, we lack any kind of organizational or classification structure even remotely similar to the periodic table of elements, which makes chemistry a deterministic science.

Any classification system of ecological units is likely to be contentious, both within the biological community and among policy makers. The controversies surrounding the identification of old-growth forests in the Pacific Northwest and the delineation of wetlands across the United States are recent examples. Others that address particular management issues are classifications of ecosystems for managing fish and forest health (Rieman et al. 2000), and the use of remote sensing data to classify forest ecosystems (Treitz and Howarth 2000). None of these examples, however, begin with a causal model of how ecosystems work in all their parts and relationships and, from that, build a classification of ecosystem types. Thus we are led to classifying ecosystems in a more operational and issue-driven manner than an actual scientific one.

But does this mean that the ecosystem concept has no value in our analyses of animal population and habitat ecology? We think not, because this concept embodies all of the interactions that influence what an animal does and the context in which lineages of animals evolve. It thus reminds us that a wild animal lives in a complex situation, and that changes in individual factors in the environment can have a cascading effect on an animal, as well as on the population to which it belongs. Any advancement in our understanding of these interactions augments our ability to manage a species. Taking such an approach forces us to view wild animals in a broader context of space and time than was previously considered in approaches to wildlife management.

How do we actually include the ecosystem concept in studies of animal ecology? Although the topic is not explicitly identified as such, much of this book has emphasized the need to organize studies of wild animals within clearly defined spatial scales and populations. Much of the research in animal ecology will, and should, remain at what generally are microscales (e.g., James and Coskey 2002), in order to be tractable, although both relatively small and large scales should eventually be integrated in an investigation where possible (e.g., Gehrt and Chelsvig 2003; Spencer and Thompson 2003).

Ultimately, viewing a population-habitat study in light of the larger ecosystem in which it is imbedded should help determine what factors should be measured, and at what scale. A rather traditional principle for understanding an ecosystem was first advanced in the general systems–theory literature of the 1960s (e.g., Lance and Williams 1967; Von Bertalanffy 1968) and later echoed in hierarchy theory (Peterson and Parker 1998). This kind of approach can be applied to studies of wild animals in an ecosystem context if we look at least 2 levels of spatial scale or organization that are smaller than that of the entity of central interest (to identify or determine the specific mechanisms responsible for observed patterns), as well as 2 levels that are larger (to describe the environmental context and emergent behaviors or patterns).

If the goal is to determine why an animal uses a specific site, the ecosystem concept tells us that it will usually be necessary to measure many things at several different scales. The ecosystem concept thus places a habitat study within a hierarchy (Kolasa 1989), not unlike the overall hierarchical concept of habitat selection. It would be helpful to objectively draw a line on a map around an ecosystem, but we do not think such a thing can happen with full consensus. It is possible, however, to draw lines among many of the factors that we know influence a wild animal, such as specific vegetation conditions, presence of cliffs and talus, and areas of moderate winter climate. Food web theory, predator-prey relationships, competitive interactions, and so on have well-developed bodies of theory (which can be used for deductive studies) and empirical research that lends insight into how animals perceive their surroundings (inductive studies). All of these interactions are part of what we consider ecosystem functions. This tells us that the days of simply measuring some plots of vegetation and calling those measurements a habitat study for a particular species (or group of species) should rapidly change into broader-based studies, although there is still a need for the former type of investigations, especially on little-known species and their habitats, examined in a context that can incorporate classic natural history (Herman 2002). Such work has formed a solid foundation upon which we can now build a better understanding of ecological relationships by expanding them beyond a simple consideration of vegetation variables as descriptors of habitat. The next big leap in studying animals in a broader setting is to identify linkages between the environmental and demographic factors that influence the trajectory of populations in an ecosystem context (e.g., Lambert et al. 2016; Kvasnes et al. 2017).

Identifying and Managing Key Environmental Correlates

Understanding, managing, and conserving populations of wild animals in an ecosystem context also raises the question of which characteristics of ecosystems to study and identify for management and planning purposes. The traditional approach to wildlife-habitat management in the past focused on the habitat elements of water, food, and cover. These are essential items for maintaining the physical health of wild animals and plants and, thus, the persistence of their populations. But this does not adequately describe all facets of ecosystems vital to ensuring realized fitness of individuals, viability of populations, and, ultimately, the persistence of a species. A broader approach has been taken in various ecological assessments of wildlife and watersheds in the western United States

(Marcot et al. 1997). In this assessment, species (and selected subspecies and plant varieties) were described by their use of key environmental correlates. These are biotic and abiotic conditions of a species' environment that proximately influence the realized fitness of individuals and the viability of populations. Such key correlates can include the biophysical attributes traditionally considered as habitat elements. They can also encompass other biotic or abiotic factors not traditionally thought to be habitat elements, such as the use of roads, air quality, hunting or collection pressure, and interspecific interactions.

The purpose of extolling the use of key environmental correlates is to extend a focus on environmental factors beyond simple descriptions of vegetation types and their structural or successional stages, which have been so commonly used for wildlife-habitat assessment and management. The use of key environmental correlates can help shed light on the effects (positive or negative) of human activities and other dynamic aspects of ecosystems beyond those affecting just vegetation conditions.

In an assessment in the interior Columbia River Basin in the western United States, Marcot et al. (1997) depicted key environmental correlates for a wide variety of selected taxa and species groups of macrofungi, lichens, bryophytes, rare vascular plants, selected soil microorganisms, arthropods, and mollusks, as well as for all vertebrates. A single hierarchical classification of key environmental correlates was developed for the organisms. A database of these correlates for each wildlife taxon was then used to determine which species shared common correlates, how the management of some correlates (e.g., large down wood in forests) affects suites of species, and what collective set of correlates should be recognized for managing the full spectrum of species in an area. This key environmental correlate classification was developed hierarchically, so groups of organisms sharing various levels of specificity for the correlates could be identified. More recently, use of such correlates has been extended to evaluate terrestrial environments as part of watersheds assessments within the entire Columbia River Basin under the aegis of the Northwest Power and Conservation Council.

Identifying Key Ecological Functions

Another facet of managing wild animals in an ecosystem context pertains to understanding the ecological roles played by different species of animals. The traditional approach to habitat management has assumed that wildlife species are a function of their habitat, and that managing wildlife simply requires providing the right kinds of habitats. A more helpful process would be to explicitly depict the array of key ecological functions (sometimes also called "key ecosystem functions") of individual species. In this context, a key ecological function (KEF) refers to the main ecological roles performed by species that influence the diversity, productivity, or sustainability of ecosystems (Marcot and Vander Heyden 2001). A particular KEF can be shared by many species, and a given species can have several key ecological functions. These can be described, along with key environmental correlates, in databases and models of species-environment relationships.

The main categories of key ecological functions (Table 6.2) include trophic relationships; herbivory; nutrient cycling; interspecies relationships; disease, pathogen, and parasite relationships; soil relationships; wood structure relationships; and water relationships. Each category, in turn, can be divided into a number of hierarchical subcategories. The KEF classification was used in a wildlife-habitat relationships database to determine which species shared specific ecological functions, as well as the array of functions performed by the species of interest. Species with the same KEF were called "ecologically functional groups" of species. Information on the key ecological functions of a species was cross-linked to data on key environmental correlates and the range distributions of that species, as discussed below. The classification of KEFs we present here may be useful for guiding other studies of wildlife biodiversity by helping investigators focus on the ecological roles of particular species,

Table 6.2.   A hierarchic classification of key ecological functions (KEFs) of plant and wildlife species

1 Trophic relationships
    1.1 heterotrophic consumer
        1.1.1 primary consumer (herbivore) (also see below, under "herbivory" in 8.2 and 8.3)
            1.1.1.1 foliovore (leaf eater)
            1.1.1.2 spermivore (seed eater)
            1.1.1.3 browser (leaf, stem eater)
            1.1.1.4 grazer (grass, forb eater)
            1.1.1.5 frugivore (fruit eater)
            1.1.1.6 sap feeder
            1.1.1.7 root feeder
            1.1.1.8 nectivore (nectar feeder)
            1.1.1.9 fungivore (fungus feeder)
            1.1.1.10 flower/bud/catkin feeder
            1.1.1.11 aquatic herbivore
            1.1.1.12 feeds in water on decomposing benthic substrate
            1.1.1.13 bark/cambium/bole feeder
        1.1.2 secondary consumer (primary predator or primary carnivore)
            1.1.2.1 invertebrate eater
                1.1.2.1.1 terrestrial invertebrates
                1.1.2.1.2 aquatic macroinvertebrates
                1.1.2.1.3 freshwater or marine zooplankton
            1.1.2.2 vertebrate eater (consumer or predator of herbivorous vertebrates)
                1.1.2.2.1 piscivore (fish eater)
            1.1.2.3 ovivore (egg eater)
        1.1.3 tertiary consumer (secondary predator or secondary carnivore)
        1.1.4 carrion feeder
        1.1.5 cannibalistic
        1.1.6 coprophage (feeds on fecal material)
        1.1.7 feeds on human garbage/refuse
            1.1.7.1 aquatic (e.g., offal and bycatch of fishing boats)
            1.1.7.2 terrestrial (e.g., landfills)
    1.2 prey relationships
        1.2.1 prey for secondary or tertiary consumer (primary or secondary predator)
2 Aids in physical transfer of substances for nutrient cycling (carbon, nitrogen, phosphorus, etc.)
3 Organismal relationships
    3.1 controls or depresses insect population peaks
    3.2 controls terrestrial vertebrate populations (through predation or displacement)
    3.3 pollination vector
    3.4 transportation of viable seeds, spores, plants, or animals
        3.4.1 disperses fungi
        3.4.2 disperses lichens
        3.4.3 disperses bryophytes, including mosses
        3.4.4 disperses insects and other invertebrates
        3.4.5 disperses seeds/fruits (through ingestion or caching)
        3.4.6 disperses vascular plants
    3.5 creates feeding, roosting, denning, or nesting opportunities for other organisms
        3.5.1 creates feeding opportunities (other than direct prey relationships)
            3.5.1.1 creates sapwells in trees
        3.5.2 creates roosting, denning, or nesting opportunities
    3.6 primary creation of structures (possibly used by other organisms)
        3.6.1 aerial structures
        3.6.2 ground structures
        3.6.3 aquatic structures
    3.7 user of structures created by other species
        3.7.1 aerial structures
        3.7.2 ground structures
        3.7.3 aquatic structures

(*continued*)

Table 6.2. *(continued)*

3.8 nest parasite
    3.8.1 interspecies parasite
    3.8.2 common interspecific host
3.9 primary cavity excavator in snags or live trees
3.10 secondary cavity user
3.11 primary burrow excavator (fossorial or underground burrows)
    3.11.1 creates large burrows (rabbit-sized or larger)
    3.11.2 creates small burrows (less than rabbit-sized)
3.12 uses burrows dug by other species (secondary burrow user)
3.13 creates runways (possibly used by other species)
3.14 uses runways created by other species
3.15 pirates food from other species
3.16 interspecific hybridization
4 Carrier, transmitter, or reservoir of vertebrate diseases
    4.1 diseases that affect humans
    4.2 diseases that affect domestic animals
    4.3 diseases that affect other wildlife species
5 Soil relationships
    5.1 physically affects (improves) soil structure, aeration (typically by digging)
    5.2 physically affects (degrades) soil structure, aeration (typically by trampling)
6 Wood structure relationships (either living or dead wood)
    6.1 physically fragments down wood
    6.2 physically fragments standing wood
7 Water relationships
    7.1 impounds water by creating diversions or dams
    7.2 creates ponds or wetlands through wallowing
8 Vegetation structure and composition relationships
    8.1 creates standing dead trees (snags)
    8.2 herbivory on trees or shrubs that may alter vegetation structure and composition (browsers)
    8.3 herbivory on grasses or forbs that may alter vegetation structure and composition (grazers)

Source: Marcot and Vander Heyden (2001)

as a complement to the more traditional assessments of biodiversity as species richness.

A species' key ecological function may often appear in more than a single ecosystem. For example, consider the ecological function of a parasite carrier or transmitter (KEF 4.3 in Table 6.2). Species with this ecological function in the inland west of the United States include least bitterns (*Ixobrychus exilis*), which are a host for ecto- and endoparasites; western sage-grouse (*Centrocercus urophasianus phaios*), which are a host for protozoan, helminth, and bacterial parasites; and snowshoe hares (*Lepus americanus*), which support a variety of ecto- and endoparasites and are a reservoir for several viruses and bacterial pathogens. The collective set of key environmental correlates—including vegetation cover types, structural stages, and other environmental

factors—that are occupied by even this simple ecological functional group of 3 species spans terrestrial forest, shrubland, grassland, and riparian (wetland) ecological systems. In this way, ecological processes for each subsystem can be identified.

Ecological processes, as used here, are those groups of key ecological functions of species that pertain to each part of the ecosystem. For example, ecological processes associated with soil subsystems include organic matter decomposition, nutrient pooling and cycling, and the provision of conditions for mesoinvertebrates and fungi that are critical to vascular plant productivity. Species ecological functions associated with such processes in soil subsystems include soil aeration, the turnover of soil nutrients and layers, nitrogen retention and uptake, and soil stabilization. With a database on KEFs, one can then determine which species play these functional roles that collectively contribute to particular ecological processes.

Further, counting the number of wildlife species with particular key ecological functions is a crude but useful measure of functional redundancy of that KEF category. The greater the functional redundancy of an ecosystem, the greater the potential resilience of that ecosystem to perturbations (Lawton and Brown 1994; Mooney et al. 1996; Gunderson et al. 2002). Figure 6.8 provides an example of a functional profile for all forest mammal species in Washington and Oregon (Marcot and Aubry 2003). The profile revealed that functional redundancy among this set of forest mammals was greatest for such functions as serving as prey for predators, consuming invertebrates, and dispersing seeds and fruits. In their analysis of mammals in western US forests, Marcot and Aubry (2003) also determined that this

assemblage of forest mammals was not particularly species rich, compared with other taxa, but it was functionally rich in terms of the collective types and numbers of key ecological functions they perform.

A species' key ecological functions and ecological processes all contribute to diversity, sustainability, and productivity over time. For example, the ecological processes and a species' KEFs associated with the soil subsystem of forest ecosystems all contribute to the biodiversity of fungi, lichens, plants, mesoinvertebrates (e.g., soil mites), macroinvertebrates (e.g., earthworms), and fossorial vertebrates (e.g., pocket gophers); the productivity of plant and animal populations, including tree growth; and the sustainability of resource growth and use (e.g., sustained timber production and harvest) over the long term.

Beyond functional redundancy, a number of species, community, and geographic patterns of key ecological functions in wildlife communities can be described (Marcot and Vander Heyden 2001). Marcot and Aubry (2003) used these categories of patterns to assess the functional roles of forest mammals. Previously, Marcot (2002) had used them to evaluate the ecological roles of vertebrate wildlife associated with decaying wood as an extension of traditional guidelines for managing snags and down wood in public forests. The aim of these evaluations of the functional roles of wild animals was to challenge managers to think functionally and include considerations for such effects in their management guidelines.

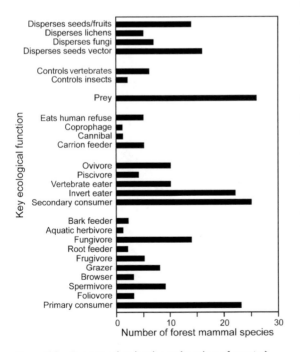

Figure 6.8   Functional redundancy (number of species) of forest mammals in Washington and Oregon, by the selected category of their key ecological function (see Table 6.2). Such functional profiles can be further specified for particular vegetation types or geographic locations. (Adapted from Marcot and Aubry (2003:639))

Functional Analysis as Part of Wildlife Assessment and Management

In general wildlife-habitat assessments, we advocate evaluating each aspect of the triad of habitats, species, and functions. In many wildlife-habitat relationship models (see Chapter 5), elements of this triad are presumed to be functions of one another, the most traditional assumption being that species are a function of habitat. Nevertheless, patterns of species richness or occurrence by community, or of an

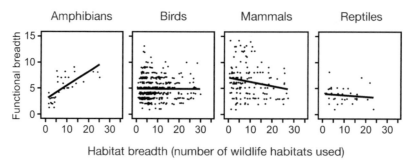

Figure 6.9   Some patterns indicating the functional aspects of wildlife do not necessarily correlate with patterns of habitat use. This composite graph plots functional breadth (number of categories of key ecological functions; see Table 6.2) against habitat breadth (number of macrohabitat types used) for all 595 species of non-fish vertebrates regularly occurring in the Columbia River Basin that are represented in a wildlife-habitat relationships matrix database (unpublished data from T. O'Neil, pers. comm.). For the most part, there is no statistically significant correlation (P > 0.05) between functional breadth and habitat breadth. The significant positive correlation with salamanders is due to a set of 16 mostly insectivorous salamanders (most or all species of Dicamptontidae, Rhyacotritonidae, and Plethodontidae) that are tied to mesic and humid forest types. The overall conclusion for wildlife assessments is to evaluate habitats, species, *and* functions separately.

individual species' use of various habitat conditions, do not necessarily correlate with their ecological functions (e.g., Fig. 6.9). Thus, each member of this triad tells a complementary story.

Various patterns of key ecological functions among species and within communities have been described (Table 6.3) and used in various projects (e.g., Marcot and Aubry 2003). Further, many of the functional patterns listed in Table 6.3 supersede the more general (and vague) concepts of keystone species, key habitats, fully functional ecosystems, and properly functioning conditions.

For example, a species-level KEF pattern that is useful for identifying a focal species or a species potentially at risk is that of a functional specialist. This is a species that performs very few ecological roles in its ecosystem (that is, a species with very few key ecological functions, as listed in Table 6.2), even if other species might also perform those roles. In western North America, functional specialists include some strictly insect- or invertebrate-feeding amphibians, such as Larch Mountain salamanders (*Plethodon larselii*) and reptiles, such as sharptail snakes (*Contia tenuis*); some insectivorous small mammals, such as Baird's shrews (*Sorex bairdi*); and

some carrion feeders, such as turkey vultures (*Cathartes aura*). Functional specialists may be very vulnerable to changes in their key resources and thus may need special conservation attention. This has been shown by the severe decline in several Old World vulture species, whose plummeting populations have been caused by either an avian disease or, as has been more recently discovered, by their carrion prey being inadvertently tainted by pharmaceutical chemicals (Margalida et al. 2014; Cuthbert et al. 2016).

Another functional pattern of use in wildlife evaluations and management in an ecosystem context is that of critical functional link species, which are the only species in a community that perform a particular ecological role. Loss of such species in a particular area (or full-blown extinction) means the loss of that function. Examples of critical functional link species are hippopotamuses (*Hippopotamus amphibius*), American bison (*Bison bison*), and American alligators (*Alligator mississippiensis*). Each of these species performs the critical function of creating or maintaining wetland pools by wallowing, and they are usually the only species in a particular area with this function. Many other species use and depend on the wetlands

Table 6.3. A taxonomy of patterns of key ecological functions (KEFs) of wildlife species and communities and how to evaluate them using a wildlife-habitat relationships database. Many of these categories are unstudied in wildlife communities; thus ecological implications should be viewed as working hypotheses. See Table 6.2 for KEF categories.

Functional pattern	Definition	Ecological implications	How to evaluate
		Community patterns	
Functional redundancy	The number of species performing the same ecological function in a community.	High redundancy imparts greater resistance of the community to changes in its overall functional integrity. Low redundancy suggests critical functions to watch.	Tally number of species by KEF category for specific wildlife habitats, comparing changes over time or among habitats.
Functional richness	The total number of KEF categories in a community.	Denotes degree of functional complexity; greater functional richness means more functionally diverse systems. Also denotes the degree to which the full functional web of a community would be provided or conserved.	Tally number of KEF categories among all species present in a wildlife habitat and, optionally, habitat structure. Compare such tallies resulting from changes in habitats and structures.
Total functional diversity	The total array of KEF categories weighted by their redundancy, i.e., the number of functions times the mean functional redundancy across all functions.	Denotes total functional capacity of a community. High functional diversity means many functions and even redundancy among functions. Low functional diversity means few functions and skewed redundancy (i.e., some functions with few species).	Tally number of species by KEF category for a given wildlife habitat. Calculate mean number of species per KEF category. Can assign weights to some KEF categories if they pertain to specific management objectives.
Functional web	The set of all KEFs within a community and their connections among species and, hence, to habitat elements.	Depicts how habitat elements provide for species and the array of ecological functions performed by those species. Functions typically extend well beyond the specific habitat elements.	Identify habitat elements of management interest. List all species within a wildlife habitat that are associated with those habitat elements. List all KEF categories associated with those species. Compare changes in habitat elements.
Functional profile	The degree of functional redundancy, compared across communities.	Identifies communities with low (or high) redundancy of particular KEF categories. This can help prioritize habitat management (e.g., to ensure continuance of low-redundancy functions).	Tally number of species by KEF category and by wildlife habitat. Identify wildlife habitats with lowest tallies for each KEF category.
Functional homology	The functional similarity of communities, even if species composition differs.	Determines if two communities are functionally homologous (i.e., if they have similar functional profiles and patterns of functional redundancy, even if the species performing the functions differ). Functionally homologous communities can be expected to operate in similar ecological ways.	Produce functional profiles across all KEF categories for several communities or for a community over time, based on its expected changes in habitat elements, habitat structures, and the like. Compare the profiles (e.g., via contingency analysis) and identify statistically similar (functionally homologous) communities.

(*continued*)

Table 6.3. (*continued*)

Functional pattern	Definition	Ecological implications	How to evaluate
Geographic patterns			
Functional bottleneck, or cold spot	Geographic locations with very low functional redundancy of an otherwise widely distributed functional category.	Denotes areas of higher risk in severing functionally connected communities across the landscape. Severing functions might set the stage for degradation of functional ecosystems.	Map wildlife habitats and distributional ranges of wildlife species. For a given KEF category, map the number of wildlife species in each habitat or overlay their range maps. Identify locations with lowest species richness bordering higher richness on each side These are geographic functional bottlenecks (cold spots).
Functional linkage, or hot spot	Geographic locations with very high functional redundancy.	Denotes areas where many species provide a specific ecological function. Such communities may be more resilient to changes in environment or habitat for that function.	Map species richness for a particular KEF category as above. Identify locations with highest species richness. These are functional linkages (hot spots). Determine which species occur in a given hot spot and their wildlife habitats, habitat elements, and habitat structures, as well as how changes might influence the persistence of the species and, thus, the redundancy of the function.
Species' functional roles			
Functional keystone species, critical functional link species, and critical functions	Functional keystone species are species whose removal would most alter the structure or function of the community. One form of this may be critical functional links, which are species that are the only ones performing a specific ecological function in a community. A critical function therefore is the associated functional category represented by only 1 (or very few) species within a community.	Reduction or extirpation of populations of functional keystone species and critical functional links may have a ripple effect in their ecosystem, causing unexpected or undue changes in biodiversity, biotic processes, and the functional web of a community.	For a given wildlife habitat, tally the number of species for each KEF category (functional redundancy). For KEF categories with only 1 species, determine which species performs this function. This is a critical functional link species for this particular function in this habitat.
Functional breadth and functional specialization of species	The number of ecological functions performed by a species.	Species with the narrowest functional breadth (i.e., fewest functions) are functional specialists and may be more vulnerable to extirpation from changes in conditions supporting that function.	For a given wildlife habitat, tally the number of KEF categories for each species. Identify the species with the fewest number of categories; these are functional specialists. Determine their habitat elements and structures and, thus, their potential vulnerability to changes thereof. Functional specialists that are also functional keystones may be of high priority for conservation attention.

Functional pattern	Definition	Ecological implications	How to evaluate
		Functional responses of communities	
Functional resilience	The capacity of a community to return to a starting pattern of total functional diversity, richness, and redundancy following a disturbance event.	Functionally resilient communities are better able to maintain their biotic processes in the face of disturbances. Conversely, it is important to know how far a community can be changed by some anthropogenic disturbance event and still be able to return to its starting functional pattern.	Determine the total functional diversity, functional richness, and functional redundancy of a pre-disturbance community. Then determine the types and rates of recovery of its wildlife habitats, habitat elements, and habitat structures following some disturbance; the wildlife species associated with such recovery stages; and the species' KEF categories and functional diversity, richness, and redundancy for each recovery stage. Compare stages for functional similarity and, thus, resilience.
Functional resistance	The ability of a community to resist changes to its functional diversity, richness, and redundancy following a disturbance event.	Functionally resistant communities can be counted on to continue to provide specific ecological functions in spite of and during disturbances. They may provide a bastion for a specific desired function in a disturbed or managed landscape.	Analyze as above and determine the degree to which functional diversity, richness, and redundancy do not change for each post-disturbance stage. This is a measure of functional resistance.
Functional attenuation	The degree to which the set of ecological functions within a community simplify following a disturbance event.	Functionally attenuated communities provide fewer or lower redundancies of ecological functions. It may be particularly important to know the degree of functional attenuation to be expected following anthropogenic disturbances.	Analyze as above and determine which KEF categories are likely drop out and which remain over post-disturbance recovery stages, as compared with initial conditions. Calculate functional diversity, richness, and redundancy for each stage to determine the rate of functional attenuation. Compare final stage to initial conditions to determine overall functional attenuation.
Functional shifting	The degree to which a community changes to a new, stable, functional constitution following a disturbance event.	Communities with low functional resilience or resistance may end up with a new array of functions and a new pattern of functional diversity, richness, and redundancy. It may be particularly important to know how a community might functionally shift following anthropogenic disturbances, and, thus, which functions might be weakened, strengthened, lost, or gained.	Analyze as above and compare KEF categories and functional diversity, richness, and redundancy between pre-disturbance and final stable stages. Identify which functions are lost or gained, and which change in redundancy.

(continued)

Table 6.3. *(continued)*

Functional pattern	Definition	Ecological implications	How to evaluate
Imperiled function	A function that is represented by very few species (i.e., critical functional link species) or by species that are themselves scarce, declining, or moribund, where extirpation of the species would mean loss of the function. Imperiled functions also can be identified geographically.	Loss of imperiled functions serves to degrade overall ecosystem integrity. Even seldom-performed ecological functions might be critical to maintaining ecosystems, such as occasional dispersal of plant seeds in the face of shifting climates.	For a given wildlife habitat, determine KEF categories with the lowest functional redundancy, as well as the risk level of the associated species. Imperiled functions are those with 1 or few species that are themselves at risk.

Source: Marcot and Vander Heyden (2001)

and ponds these species create or maintain. The removal of a critical functional link species would eliminate their function and probably adversely affect many other species (e.g., Campbell et al. 1994; Knapp et al. 1999; Matlack et al. 2002). The many other function-related patterns listed in Table 6.3 likewise can be and are being used in wildlife evaluations to help set priorities for wildlife-habitat management.

An Ecography of Key Ecological Functions

The concept of key ecological functions can be expanded to a practical geographic context. Using GIS, maps of functional redundancy have been produced for a variety of projects in the western United States (e.g., Marcot et al. 2003). Such maps are constructed by stacking the maps of known or expected occurrences (such as from habitat relationships) for those species with specific key ecological functions. An example is shown in Figure 6.10, which maps the changes in functional redundancy from historic to current time periods for the soil-digging function. Soil digging by vertebrate (as well as invertebrate) organisms can increase soil porosity and the uptake of organic matter, affect the bulk density of soils, and otherwise variously influence the productivity of soils in an area (Scheu et al. 2002; Canals et al. 2003). The map in Figure 6.10 shows how the redundancy of this function has declined significantly in many of the major valley and inland lowland areas of the Pacific Northwest—such as the Willamette Valley in western Oregon, the Columbia Basin in southeastern Washington, and the Snake River Valley in southern Idaho—largely because of the agricultural conversion of native grasslands, shrublands, steppes, and woodlands in these areas. Restoration of this ecological function may entail the rehabilitation of some of these native environments, along with the native species that perform this function. In this way, geographic patterns of key ecological functions and their changes can be mapped and used in broad-scale planning and assessment efforts.

Beyond Habitat: Species-Environment Relationships

A relational database can list wildlife species and their key environmental correlates, key ecological functions, and range distributions. It extends beyond the traditional wildlife-habitat relationship approach by considering non-habitat environmental elements and species' ecological functions. With queries devised to look at issues spanning species and their correlates, functions, and distributions, such a database can be termed a more general "species-environment relationship model." This model can

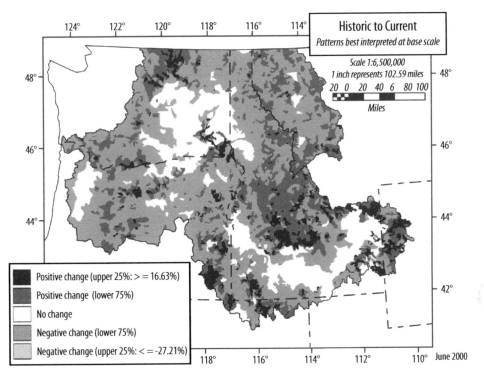

Figure 6.10   Changes in functional redundancy (the number of all terrestrial vertebrate wildlife species) from historic (early 1800s) to current (ca. 2000) time periods, for the key ecological function (KEF) of soil digging (see KEF 5.1 in Table 6.2). Such maps display locations where functions are likely to have declined or increased. This map was produced by intersecting maps of wildlife-habitat types with information on species occurrence by habitat type and their KEF categories. (Reproduced from Marcot and Aubry (2003:642))

be used to pose new questions about the roles of species in ecosystems and the ways in which managers might make provisions for species ecological functions through habitat management. For instance, which species provide specific ecological functions, such as nutrient cycling, soil turnover, or insect population control? What are the collective habitat and environmental requirements of such species? Where do they occur? How do such ecological functions affect productivity and diversity of ecosystems? Such questions are now answerable and being used in wildlife assessment projects, as illustrated by the examples given in the previous section.

An investigator could delineate an ecological function and identify all species having this function—along with their collective set of habitat requirements and ecological correlates—by subsystem. This would

help identify the set of conditions that managers might provide to help maintain habitat conditions for species with designated ecological functions. Other approaches are also possible, such as overlay mapping for all species associated with a particular ecological function, as a way to represent the full geographic distribution of (and potential management risks to) specific functions. In a sense, this would extend the current approaches of spatially explicit modeling of wildlife populations to include species' ecological functions. The classification of land units, including ecoregions or bioregions and existing or potential vegetation communities, are useful and often essential for conservation assessments, but, by themselves, they do not tell us much about the status and trends of associated species of wild animals or their populations.

Ecosystem Services

Key ecological functions and key environmental correlates of wildlife animals, and the environments in which species occur, can be further interpreted in terms of the ecosystem services (or ecological services) they provide. The concept of "ecosystem services" refers to the array of natural resources and processes that are valued by humans and sustain ecological communities and ecosystems. The literature on ecosystem services describes a very wide array of such resources and processes (e.g., Daily et al. 1997; Ostfeld and LoGiudice 2003; Crossman et al. 2013). In reviewing this literature, we developed a set of five categories of ecosystem services that can be used to focus discussion on this topic (Table 6.4). These are (1) ecological functions of organisms that maintain ecosystem integrity, such as the pollination of desirable plants and food crops; (2) biophysical services mediated by environmental processes and conditions, such as the filtering of surface water; (3) ethnobiotic services that provide cultural, social, and religious values for organisms, such as medicinal plants; (4) economic services that offer largely intangible benefits, such as recreational experiences; and (5) natural resource services that supply renewable commodities with tangible and direct economic values, such as game animals. Many other examples are also listed in Table 6.4.

The purpose of classifying ecosystem services is to better identify the many important roles that wildlife species and their ecological functions and habitats play in providing for human needs and ecological systems alike. A valuation of ecosystem services that are of interest to people can help identify the economic benefits of managing wildlife species and their habitats (Scott et al. 1998). Rudiments of the economic benefits supplied by ecosystem services have been described for decades. Such examples include the value of certain bird and bat species that consume insect pests (e.g., Takekawa and Garton 1984), and the economic value of pollinator services (Cane and Tepedino 2001). A more complete classification of ecosystem services can help investigators recognize and quantify the many economic, ecological, social, and even spiritual benefits of maintaining wildlife. There is a huge body of literature on concepts related to classifications and analyses of ecosystem services. For instance, Costanza et al. (2006) and Maes et al. (2012) offer excellent examples of ecosystem services analyses on a state scale (New Jersey) and a sub-continental scale (Europe), respectively.

Beyond Viability

Much continues to be written about whether wildlife management should focus on individual species or on ecosystems (e.g., Franklin 1993; Bowen 1999; Noon et al. 2009). We do not view this argument as being particularly fruitful, because both approaches are necessary for ensuring the long-term viability of populations and, hence, the evolutionary potential of species. A management approach that integrates habitat components with a defined population of interest and other environmental correlates, as well as with factors that explicitly relate to species' functions and ecosystem processes, should strive to encompass several levels of biological organization. Simple coarse-filter approaches to wildlife-habitat management may be a useful starting point, but they cannot account for the requirements of all ecological entities.

In a more comprehensive approach (as outlined above), the problems and complexities of defining minimum viable population levels and prioritizing individual species for recovery efforts, such as in a triage approach, would be reduced. Ultimately, we are all part of the ecological systems we try to manage, and models that describe species of wild animals can just as well be built for humans. In this way, species, ecosystems, ecological functions, environmental conditions, ecosystem services, management goals, and management activities can all be merged into a single coherent approach to their assessment and management. The utility of this comprehensive strategy should become more evident for managing

Table 6.4.   Classification of ecosystem services

Class of ecosystem service	Definition	Examples
Ecological services of organisms	Key ecological functions and services directly provided or mediated by organisms. This refers to the strictly ecological roles, played by organisms, that help keep ecosystems diverse, productive, and fully functional.	• pollination • carbon sequestration • detoxifying xenobiotics • decomposition of wastes • generation and renewal of soil and soil fertility • nitrogen mineralization • soil protection by biotic crusts • slope stabilization by roots and down wood • dispersal of seeds, spores, and disseminules • translocation of nutrients
Biophysical services	Services mediated or provided by macrohabitats in conjunction with biophysical conditions. This refers more to processes of the physical environment.	• filtering of water by wetlands • filtering of water through hyporheic zones • filtering of air by forest vegetation • provision of oxygen by plants • amelioration of fire intensity • moderation of floods and droughts • protection from the sun's harmful UV rays • partial stabilization of climate • moderation of temperature extremes and the force of winds and waves
Ethnobiotic services	Cultural values of plants and animals. This refers to cultural, religious, and other intangible values, as well as to native medicinal uses.	• medicinal plants used by native peoples • plants and animals of cultural or religious significance • existence value of organisms • provision of esthetic beauty and intellectual stimulation that enrich the human spirit
Economic services	Services of direct economic significance. These are services of economic significance but are not direct tangible commodities per se.	• pollination of crop plants • biological control of agricultural or forestry pests, diseases, or pathogens • recreational services, such as birdwatching, wildlife viewing, and ecotourism • direct use of plant products in pharmaceutical industry
Natural resource services	Direct provision of renewable natural resources and commodities of economic value. These are services that directly produce or support resources to be used or harvested and sold in the marketplace.	• timber (forest growth cycle) • clean water (hydrologic cycle) • game animals and furbearers • non-timber forest products (mosses, mushrooms, ferns, etc.) • healthy rangelands for livestock grazing

wildlife in an evolutionary context. A useful means of improving our knowledge about what works best for maintaining evolutionary potential is through an adaptive management approach.

The Adaptive Management Approach

The adaptive management approach necessitates identifying areas of scientific uncertainty, devising field management activities as real-world experiments to test that uncertainty, learning from the outcomes of such experiments, and recrafting management guidelines based on the knowledge that is gained (Holling 1978; Walters 1986; Irwin and Wigley 1993). Modeling can play a key role in formalizing our current knowledge and identifying important areas of uncertainty (Walters et al. 2000; Lynam et al. 2002). In an ideal situation, management guidelines equate to the creation of testable hypotheses (Havens and Aumen 2000); monitoring and adaptive management studies correlate with conducting the experiment; and revision of the management guidelines is connected to reevaluation and interpretation of the study results, in terms of testing the validity of the initial hypothesis.

Adaptive management, however, has seldom been applied successfully or fully in managing wildlife habitats and ecosystems (Gray 2000). Problems that result in successful applications of adaptive management are both technical and administrative (Lee and Lawrence 1986), and they need to be overcome for adaptive management projects to be successful (Westgate et al. 2012). For example, Sims et al. (2014:1153) noted that there is a vast gulf between theoretical and empirical papers on adaptive management in the literature. Less than 13 percent of the publications on adaptive management are actual applications of this approach. Nevertheless, adaptive management holds a strong potential for advancing our understanding of how management actions might impact populations of wild animals and the habitats that support them (Robbins 2012). The following are some basic tenets of how to approach adaptive management. These tenets can be used as a checklist for specific programs, in order to ensure successful application of adaptive management in an ecosystem context.

Three Tenets of a Real-World Approach to Adaptive Management

Tenet 1. *The administration of an organization implementing adaptive management must have the political will—and explicit procedures—to accept changes in the status quo of current operations.* This first tenet summarizes a great deal of experience in the administrative behavior and political reality of resource management organizations and institutions. Despite good intentions and explicit promises in management plans, if an organization lacks the political motivation or the cultural environment required for accepting new knowledge, then management activities will remain unaltered. What is needed is a clear political mandate to accept change in a timely manner, and a formal protocol for weighing and incorporating new data for potential use in updating or reaffirming management decisions.

A corollary to this tenet is that an explicit risk management framework be put into place to accept and weigh new inputs in a timely fashion. To use this information, resource management organizations must have a basis for acknowledging and incorporating uncertainty into their decision-making processes (Gunderson 1999). Often, however, such information is treated as being antithetical to carrying out the management or planning guidelines already chosen, and it is only employed under duress (e.g., in litigation) or during intermittent updates in predefined planning cycles (Clark et al. 1996). For federal resource agencies in the United States, such planning cycles are years or decades long.

Another corollary to this tenet is that performance standards for managers or decision makers must explicitly address environmental conservation. Here is the heart of the problem. If managers or decision makers are not held individually responsible for meeting the objectives of adaptive management (and wildlife-habitat management), then new information will not be used and management efforts will not change. In this context, "environmental conservation" refers to wise sustainable use. One facet of this may (but not necessarily) include the strict preservation of habitats or environments.

A third corollary is that what is at risk is the condition of the land, not the status of one's career or the acceptability of political decisions related to management directives. Risk analyses and risk management for wildlife typically assume that the very concept of what is at risk is shared by biologists, managers, decision makers, and politicians alike (e.g., Graham et al. 1991). In reality, this is far from the truth. Biologists explicitly speak of risk in the sense of the likelihood of species extirpation. Managers and decision makers may implicitly refer to the concept of risk, but they do so more in reference to how a decision may impact their career status or the accomplishment of overall organizational directives that may extend beyond species or habitat objectives. Upper-management decision makers and politicians may view risk in terms of their political decision space—what is politically acceptable to their peers, to special interest groups, to funding sources, or to their voting constituency. Each of these aspects of risk is authentic but quite different. Only through clearly specified performance standards for meeting adaptive management goals can specialists, managers, and decision makers be assured of viewing the concept of risk in the same way. In the absence of such standards, the next best approach is to clearly articulate the basis (or bases) for risk in a given assessment or decision, so it is clear to all how the term and concepts were defined and weighed in any decision affecting public lands and wildlife.

Use of decision modeling techniques can aid managers in choosing an optimal course of action and articulating their decision criteria. For example, the decision model of McNay et al. (1987) provided a means of prioritizing management for deer populations in coastal British Columbia in Canada, as well as for explicitly stating the environmental and management conditions that formed the basis for these priorities. Such models, however, are of limited utility if the decision makers do not wish to follow such a rigorous procedure or expose their decision-making criteria, because of its political risk.

Tenet 2. *Options exist to implement change.* This may seem to be an obvious point, but it is often overlooked or deliberately not addressed. In some cases, extensive funding has been provided for research or monitoring while options have vanished for changing conditions on the land, such as through the protection or restoration of dwindling habitats. One example may be that of Mt. Graham red squirrels (*Tamiasciuris hudsonicus grahamensis*), a potentially threatened subspecies occurring on a mountaintop in the southwestern United States, a locale that is also coveted as a development site for an astronomical observatory (Warshall 1995). In this instance, ongoing population studies are being conducted during the initial development of the observatory site. This stage is designed to retain options for either more- or less-restrictive habitat conservation, depending on the findings, although the degree of risk to the squirrel population is still subject to some debate. In other cases, however, monitoring may proceed while scarce habitat continues to be changed, such as what occurred for populations of endangered Lanyu scops owls (*Otus elegans botolensis*) on an island off of Taiwan (Severinghaus 1992), while tropical forests there were felled or converted to other uses.

An adaptive management approach needs to determine the rate of change in habitats, species, or populations of dire concern and compare this with the pace of the information-gathering and decision-making (or decision-changing) process, to ensure that critical conservation options are not lost during this initial stage. More fundamentally, conducting monitoring and research studies should not be used in place of making difficult decisions about the allocation of scarce resources, or as a smokescreen to permit the continuance of management activities that eliminate options for conservation of a scarce and dwindling resource. Monitoring, research, and funding all support—but do not substitute for—sound resource management decisions and actions.

A corollary to this tenet is to ensure that irreversible losses of resources or environmental conditions are not incurred during the testing period. For example, monitoring the population dynamics or demography of an endangered species while its habitat continues to be adversely altered violates this corollary and assuredly guarantees serious problems for meeting conservation goals. In some cases, adaptive management experiments can deliberately sacrifice or seriously alter some environmental conditions for a habitat or species of interest, albeit doing so in the name of quantifying the effects of management activities. Yet the pace of such experiments and the degree of reversibility of any losses must still permit options for changing management activities, such as for the in situ protection or recovery of threatened species.

Another corollary is that changes in environmental conditions due to human activities (ongoing management or new experiments), together with those from natural disturbances, do not outstrip the pace of monitoring and learning, as well as the potential to alter management activities. In other words, adaptive management approaches should attend to the additive effects from both human and natural causes, and then gauge the likelihoods of being able to change activities in time to ensure that they meet conservation objectives. For example, if management experiments are testing the effect of various degrees of draining in some rare wetland type over a specified amount of time, the likelihood of environmental catastrophes, such as a prolonged drought during the same period, should be factored in to help ensure that options for change still exist at the end of the experiment.

Tenet 3. *Indicator variables—environmental parameters—can be identified and realistically monitored in a cost-effective way.* Many adaptive management studies may be well designed in theory but cannot be realistically carried out. For instance, an investiga-

tion might require indicator variables that cannot be readily identified or measured, or the experiment might be so complex or costly that it could not be completed with adequate sample sizes or be sufficiently intense. This is often what happens in studies designed to survey the response of carnivore populations to management activities, for example, or examinations of ecological processes across entire landscapes, which cannot be replicated. If this proves to be the case, then the management directives and the focus for associated adaptive management studies need to be more tightly specified, and new statistical approaches should be considered (e.g., Reckhow 1990).

Another corollary to this tenet is that a statistical sampling frame should be established, in order to distinguish the effects of human activities from background changes. One premise of adaptive management is that we are able to differentiate alterations caused by human activities from those caused by natural variations or background noise. This can be determined through carefully designed manipulation experiments—and far less often by using "natural experiments" or passive observation studies (for a discussion of appropriate study designs, see Chapter 1). This is vital for judging when we can and when we cannot affect change through management activities. Adaptive management studies need to pay close attention to sampling design and the use of controls and treatments, as well as to ensure specific power and confidence levels in statistical tests.

Conservation questions often pertain to conditions that occur across very large landscape areas and, thus, in contexts that cannot be replicated with adequate controls and in sufficient numbers to meet the assumptions of traditional statistical techniques. We therefore must look to other statistical designs, including the use of comparative time series and spectral analysis (de Valpine and Hastings 2002), empirical Bayesian statistics (Oman 2000), and optimization approaches (Hof and Bevers 1999), although some of these methods are controversial. In certain cases, entire landscapes can be devoted to demonstration experiments (Lindenmayer et al. 1999). We advocate that a plurality of approaches be taken to provide the widest possible means of learning when unique conditions violate the assumptions of traditional statistics.

A third corollary is that the research objectives and expected effects are clearly articulated and quantified. This may seem obvious, but it is often overlooked, particularly in observational (non-experimental) studies that result in post hoc "fishing expeditions" for patterns and management affects, and in demonstration studies designed mostly to justify and then conduct a priori desired activities (Guthery 2008).

Overall, we make the following suggestions for helping to ensure a successful approach to the adaptive management of habitats for wildlife. First, review the above tenets and their corollaries and make a checklist. Second, clearly separate risk analysis from risk management. Risk analysis should be based on estimating (and partitioning) the likelihoods of outcomes resulting from potential management actions and those from natural conditions and changes (or chance events). Risk management should incorporate and articulate the criteria used in reaching a management decision, including explicating its risks in light of uncertain projections and incomplete information.

Third, in the future we will need to learn from all kinds of information-gathering approaches, including observational or correlative studies, controlled field experiments, laboratory experiments, uncontrolled field trials, and anecdotal experience, as well as theoretical models. All of these approaches can be useful and should be complementary in an adaptive management framework. The challenge is in appropriately combining information gathered from disparate studies and approaches into an overall understanding of the real-world system.

Fourth, technical staffs can help managers interpret and understand the ramifications of uncertainties and the probabilities of outcomes. Many resource decision makers may not be particularly adept in dealing with scientific uncertainty (Policansky 1993). For example, they may view scientific uncertainty as lack of "proof" of any particular effect and thus believe that it can be discounted and ignored, or even touted as supporting evidence, in management decisions. This is not an appropriate interpretation of scientific uncertainty. And it is not a correct understanding of the scientific process. To verify a hypothesis, we do not seek proof, but rather corroboration or, more accurately, a lack of statistical falsification.

Also—and this is greatly misunderstood by many policy makers—uncertainty is not the same as a complete lack of knowledge. An uncertain outcome can be expressed as a low likelihood or a high variation in potential outcomes, whereas a lack of knowledge simply means that we do not know and cannot estimate outcome likelihoods. That is, lack of knowledge is not the same as high variance. Ecosystems and their component communities, populations, processes, and disturbance regimes can fluctuate highly through space and over time, and this variance can be precisely measured. In a sense, we can be certain of high variance in more than a few systems (for examples, see Wiens 2016). What may be uncertain is a specific future population level or resource productivity level (e.g., the population of some game species in a stated future decade). Even this may be expressed as an expected level with some degree of associated variance (such as a mean herd size of harvestable bull elk, plus or minus a standard error of prediction). A decided improvement would be to refer to inventories based on samples, and predictions based on projections, as "likelihoods" and "variances." This would help managers and decision makers better understand and interpret uncertainty and, where appropriate, a particular lack of knowledge. As a final point about adaptive management, Armitage et al. (2009) argued that in dealing with issues of social and ecological complexity, what we really need to implement is adaptive co-management. The term "co-management" simply means embracing strategies that bolster collaboration and learning among stakeholders and agencies, in order to foster trust building among networks of researchers, communities, and policy makers.

Real-World Circumstances and the Use of Complementary Studies

Ideally, adaptive management would proceed as rigorously defined experiments adhering to all the assumptions and tenets of well-designed scientific studies. In reality, such studies have to contend with a number of problems, including

- landscape-scale studies with few or, more usually, no replicates;
- experiments with no controls or with controls in vastly different situations (e.g., higher-elevation wilderness areas);
- little or no time to collect baseline data;
- losses of selected samples and declining sample sizes over the course of the study, because of changes in administrative or management direction;
- unannounced and undirected treatment of controls;
- overall short duration, with few truly long-term studies to determine lag and secondary effects; and
- changes in management objectives, treatments, and sometimes even land ownership over the course of the study.

What can be done in the face of such potentially ruinous circumstances? The answer may be found in using multiple studies of various kinds, as well as in taking advantage of prior knowledge to establish study objectives, management activities to test, and analysis of the results. In some cases, the real-world circumstances listed above degrade an otherwise rigorously defined experiment to the status of an observational or demonstration study. Such studies can still have value, though, by incrementally adding evidence to help corroborate or refute management hypotheses. In the southeastern United States, observational data from species of mammals (Fig. 6.11, A) and birds (Fig. 6.12, A) associated with shortleaf pine–bluestem (*Pinus echinata–Andropogon*) forests in the absence of prescribed fire indicated that a 1–5 year prescribed fire return interval has a massive influence on the composition of the vertebrate community. For mammals (Fig. 6.11, B), it took the combined results of 5 studies for this pattern to emerge (Atkeson and Johnson 1979; Tappe et al. 1999, 2004; Masters et al. 1998, 2002). For birds (Fig. 6.12, B), it took the combination of 6 studies to reveal this pattern (Johnston and Odum 1956; Meyers and Johnson 1978; Engstrom et al. 1984; Jennelle 2000; Masters et al. 2002). Although the data presented in Figures 6.11 and 6.12 only represent the presence or absence of species in relation to the structure of forest vegetation, these examples represent excellent first approximations of animal-habitat relationships that can be used to generate further investigations of demography and population performance of selected vertebrate species in this system.

The outcomes of observational or demonstration studies should not, however, be taken as hard evidence of the correctness of assumptions underlying a management approach in the absence of rigorous supporting investigations. It is just too easy to find situations and craft (inadvertently or otherwise) observational or demonstration studies to provide specific answers, regardless of actual effects. One example is locating species thought to be closely associated with specific environments in situations other than what would truly be the modal condition of the population, such as observing a few errant spotted owls in very young-growth landscapes. This is not proof (or corroboration) of such a habitat requirement. Again, a situation like this points to the need to integrate demographic data for a population with specific habitat conditions.

Retrospective studies are another kind of investigation that can greatly complement the adaptive management approach. The use of such studies is

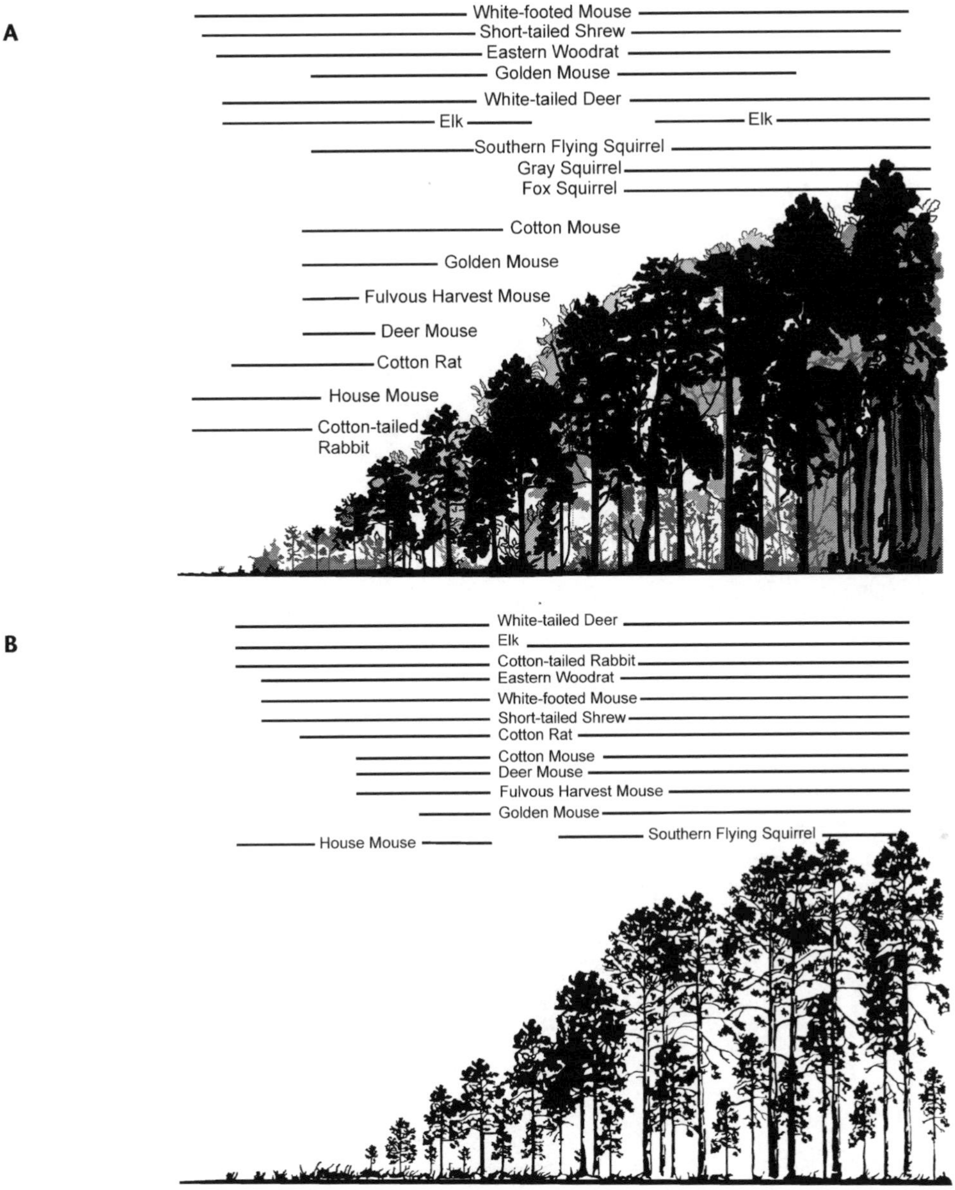

A

White-footed Mouse
Short-tailed Shrew
Eastern Woodrat
Golden Mouse
White-tailed Deer
Elk   Elk
Southern Flying Squirrel
Gray Squirrel
Fox Squirrel
Cotton Mouse
Golden Mouse
Fulvous Harvest Mouse
Deer Mouse
Cotton Rat
House Mouse
Cotton-tailed Rabbit

B

White-tailed Deer
Elk
Cotton-tailed Rabbit
Eastern Woodrat
White-footed Mouse
Short-tailed Shrew
Cotton Rat
Cotton Mouse
Deer Mouse
Fulvous Harvest Mouse
Golden Mouse
House Mouse   Southern Flying Squirrel

Figure 6.11 (*A*) A plant succession and mammal community succession model of the occurrence of some common species associated with different stages of succession within shortleaf pine–bluestem (*Pinus echinata–Andropogon*) forests in the absence of prescribed fire. (*B*) A similar model of plant succession and common species associated with different stages of succession within shortleaf pine–bluestem forests with frequent prescribed fire return intervals, ranging from 1 to 5 years. (Adapted from Block and Conner (2016), based on Atkeson and Johnson (1979), Tappe et al. (1994, 2004), and Masters et al. (1998, 2002))

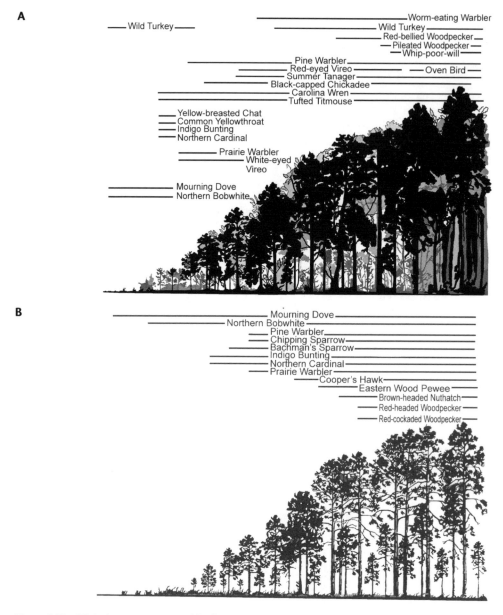

Figure 6.12 (A) A plant succession and bird community succession model of the occurrence of some common species associated with different stages of succession within shortleaf pine–bluestem (*Pinus echinata–Andropogon*) forests in the absence of prescribed fire. (B) A similar model of plant succession and common species associated with different stages of succession within shortleaf pine–bluestem forests with frequent prescribed fire return intervals, ranging from 1 to 5 years. (Adapted from Block and Conner (2016), based on Johnston and Odum (1956), Meyers and Johnson (1978), Engstrom et al. (1984), Wilson et al. (1995), Jennelle (2000), and Masters et al. (2002))

often of great value for understanding the developmental history and past conditions of habitats and wildlife. Retrospective studies are not experiments. Rather, they are hindcast reconstructions of prehistoric or historic events and conditions, in order to help us better understand long-term or recent changes. These studies can borrow from a variety of tools and techniques, including dendrochronology, palynology, paleoecology, archaeology, analysis of historic documents, use of historic photo points, and other information sources or methods.

Retrospective studies can also help inform us about long-term changes in environments, climates, and biota, as well as complex effects of land management actions (e.g., Caswell 2000; Louda et al. 2003). For instance, Crumley (1993:377) provided examples of retrospective studies to unravel "complex chains of mutual causation in human-environment relations" by tracing past human-environment interactions at global, regional, and local scales. (Compare this with the discussion on scales of ecological disciplines and management issues that has been presented in Chapter 5). Covington et al. (1994) analyzed historic alterations in forest ecosystems in the inland west of the United States to help project future changes. Studies that integrate phylogeography (study of the historical processes that may be responsible for the contemporary geographic distributions of a species or group of species) and ecological niche models (Williford et al. 2016) can provide a foundation for projecting future changes in the geographic distribution of animals, based on shifting climatic variables, landscape metrics, or both.

Summary

The traditional approach to animal ecology and conservation has been extremely vertebrate centric. Consider, for example, much of the material in this book. We urge our community of wildlife scientists, animal ecologists, and managers to reexamine our species and life-form biases and open the doors to include as many organisms and environments as possible in our work. In this way, we envision a future era of animal ecology and wildlife management where research and management efforts equally weigh the host of mostly unknown creatures—including soil, aquatic, and canopy microbes and mesoinvertebrates, as well as cryptogams and vascular plants—both for their own sakes and for their influences on vertebrates. Collectively, all of these organisms are critical to the conservation of the vertebrates, forests, grasslands, and wetlands we strive to sustain.

LITERATURE CITED

Almpanidou, V., G. Schofield, A. S. Kallimanis, O. Türkozan, G. C. Hays, and A. D. Mazaris. 2016. Using climatic suitability thresholds to identify past, present, and future population viability. Ecological Indicators 71:551–556.

Alström, P., P. C. Rasmussen, C. Zhao, J. Xu, S. Dalvi, T. Cai, Y. Guan, R. Zhang, M. V. Kalyakin, F. Lei, and U. Olsson. 2016. Integrative taxonomy of the plain-backed thrush (Zoothera mollissima) complex (Aves, Turdidae) reveals cryptic species, including a new species. Avian Research 7(1):1–39, doi:10.1186/s40657-016-0037-2.

Anthony, C. D., M. D. Venesky, and C.-A. Hickerson. 2008. Ecological separation in a polymorphic terrestrial salamander. Journal of Animal Ecology 77:646–653.

Arbogast, B. S., K. I. Schumacher, N. J. Kerhoulas, A. L. Bidlack, J. A. Cook, and G. J. Kenagy. 2017. Genetic data reveal a cryptic species of New World flying squirrel: Glaucomys oregonensis. Journal of Mammalogy 98(4):1027–1041, doi:10.1093/jmammal/gyx055.

Armitage, D. R., R. Plummer, F. Berkes, R. I. Arthur. A. T. Charles, I. J. Davidson-Hunt, A. P. Diduck, N. C. Doubleday, D. S. Johnson, M. Marschke, P. McConney, E. W. Pinkerton, and E. K. Wollenberg. 2009. Adaptive co-management for ecological complexity. Frontiers in Ecology and the Environment 7:95–102.

Arponen, A. 2012. Prioritizing species for conservation planning. Biodiversity and Conservation 21:875–893.

Atkeson, T. D., and A. S. Johnson. 1979. Succession of small mammals on pine plantations in the Georgia Piedmont. American Midland Naturalist 101:385–392.

Avise, J. C. 1995. Mitochondrial DNA polymorphism and a connection between genetics and demography of relevance to conservation. Conservation Biology 9:686–690.

Baltensperger, A. P., J. M. Morton, and F. Huettmann. 2017. Expansion of American marten (Martes americana) distribution in response to climate and landscape change on the Kenai Peninsula, Alaska. Journal of Mammalogy 98(3):703–714.

Bartelt, P. E., R. W. Klaver, and W. P. Porter. 2010. Modeling amphibian energetics, habitat suitability, and movements of western toads, *Anaxyrus* (= *Bufo*) *boreas*, across present and future landscapes. Ecological Modelling 221:2675–2686.

Beever, E. A., J. D. Perrine, T. Rickman, M. Flores, J. P. Clark, C. Waters, S. S. Weber, B. Yardley, D. Thoma, T. Chesley-Preston, K. E. Goehring, M. Magnuson, N. Nordensten, M. Nelson, and G. H. Collins. 2016. Pika (*Ochotona princeps*) losses from two isolated regions reflect temperature and water balance, but reflect habitat area in a mainland region. Journal of Mammalogy 97:1495–1511.

Bell, D. M., and D. R. Schlaepfer. 2016. On the dangers of model complexity without ecological justification in species distribution modeling. Ecological Modelling 330:50–59.

Benson, J. F., P. J. Mahoney, T. W. Vickers, J. A. Sikich, P. Beier, S. P. D. Riley, H. B. Ernest, and W. M. Boyce. 2019. Extinction vortex dynamics of top predators isolated by urbanization. Ecological Applications 29(3):e01868.

Bland, L. M., J. A. Rowland, T. J. Regan, D. A. Keith, N. J. Murray, R. E. Lester, M. Linn, J. P. Rodríguez, and E. Nicholson. 2018. Developing a standardized definition of ecosystem collapse for risk assessment. Frontiers in Ecology and the Environment 16:29–36.

Blank, L., I. Sinai, S. Bar-David, N. Peleg, O. Segev, A. Sadeh, N. M. Kopelman, A. R. Templeton, J. Merilä, and L. Blaustein. 2013. Genetic population structure of the endangered fire salamander (*Salamandra infraimmaculata*) at the southernmost extreme of its distribution. Animal Conservation 16:412–421.

Block, W. M., and L. M. Conner. 2016. Effects of prescribed fire on wildlife and wildlife habitat in selected ecosystems of North America. Technical Review 16-01. The Wildlife Society, Bethesda, MD.

Bonn, A. A., S. L. Rodrigues, and K. J. Gaston. 2002. Threatened and endemic species: Are they good indicators of patterns of biodiversity on a national scale? Ecology Letters 5:733–741.

Bowen, B. W. 1999. Preserving genes, species, or ecosystems? Healing the fractured foundations of conservation policy. Molecular Ecology 8:S5–S10.

Butler, K. F., and T. M. Koontz. 2005. Theory into practice: Implementing ecosystem management objectives in the USDA Forest Service. Environmental Management 35(2):138–150.

Butler, R. S., and R. L. Mayden. 2003. Cryptic biodiversity. Endangered Species Bulletin 28(2):24–26.

Campbell, C., I. D. Campbell, C. B. Blyth, and J. H. McAndrews. 1994. Bison extirpation may have caused aspen expansion in western Canada. Ecography 17:360–362.

Campos, P. R. A., V. M. de Oliveira, and A. Rosas. 2010. Epistasis and environmental heterogeneity in the speciation process. Ecological Modelling 221:2546–2554.

Canals, R. M., D. J. Herman, and M. K. Firestone. 2003. How disturbance by fossorial mammals alters N cycling in a California annual grassland. Ecology 84:875–881.

Cane, J. H., and V. J. Tepedino. 2001. Causes and extent of declines among native North American invertebrate pollinators: Detection, evidence, and consequences. Conservation Ecology 5(1):1, https://www.ecologyandsociety.org/vol5/iss1/art1/.

Caswell, H. 2000. Prospective and retrospective perturbation analyses: Their roles in conservation biology. Ecology 81:619–627.

Ceballos, G., P. Rodriguez, and R. A. Medellin. 1998. Assessing conservation priorities in megadiverse Mexico: Mammalian diversity, endemicity, and endangerment. Ecological Applications 8:8–17.

Christensen, N. L., A. M. Bartuska, J. H. Brown, S. Carpenter, C. D'Antonio, R. Francis, J. F. Franklin, J. A. MacMahon, R. F. Noss, and D. J. Parsons. 1996. The report of the Ecological Society of America Committee on the Scientific Basis for Ecosystem Management. Ecological Applications 6:665–691.

Clark, T. W., R. P. Reading, and A. L. Clarke. 1996. Endangered species recovery: Finding the lessons, improving the process. Island Press, Washington, DC.

Copeland, J. P., K. S. McKelvey, K. B. Aubry, A. Landa, J. Persson, R. M. Inman, J. Krebs, E. Lofroth, H. Golden, J. R. Squires, A. Magoun, M. K. Schwartz, J. Wilmot, C. L. Copeland, R. E. Yates, I. Kojola, and R. May. 2010. The bioclimatic envelope of the wolverine (*Gulo gulo*): Do climatic constraints limit its geographic distribution? Canadian Journal of Zoology 88:233–246.

Cortner, H. J., J. C. Gordon, P. G. Risser, D. E. Teegaurden, and J. W. Thomas. 1999. Ecosystem management: Evolving model for stewardship of the nation's natural resources. Pp. 3–30 *in* R. C. Szaro, N. C. Johnson, W. T. Sexton and A. D. Malk, eds. Ecological stewardship: A common reference for ecosystem management, vol. 2. Elsevier Science, Oxford.

Costanza, R., M. A. Wilson, A. Troy, A. Voinov, and S. Liu. 2006. The value of New Jersey's ecosystem services and natural capital. New Jersey Department of Environmental Protection Contract No. SR04-075. University of Vermont, Burlington.

Covington, W. W., R. L. Everett, R. Steele, L. L. Irwin, T. A. Daer, and A. N. D. Auclair. 1994. Historical and anticipated changes in forest ecosystems of the inland west of the United States. Pp. 13–63 in R. N. Sampson, D. L. Adams, and M. J. Enzer, eds. Assessing forest ecosystem health in the inland west. Haworth Press, New York.

Crossman, N. D., B. Burkhard, S. Nedkov, L. Willemen, K. Petz, I. Palomo, E. G. Drakou, B. Martin-Lopez, K. Boyanova, R. Alkemade, B. Egoh, M. B. Dunbar, and J. Maes. 2013. A blueprint for mapping and modeling ecosystem services. Ecosystem Services 4:4–14.

Crumley, C. L. 1993. Analyzing historic ecotonal shifts. Ecological Applications 3:377–384.

Cuthbert, R. J., M. A. Taggart, M. Saini, A. Sharma, A. Das, M. D. Kulkarni, P. Deori, S. Ranade, R. N. Shringarpure, T. H. Galligan, and R. E. Green. 2016. Continuing mortality of vultures in India associated with illegal veterinary use of diclofenac and a potential threat from nimesulide. Oryx 50:104–112.

Cuyckens, G. A. E., P. G. Perovic, and M. Herrán. 2017. Living on the edge: Regional distribution and retracting range of the jaguar (Panthera onca). Animal Biodiversity and Conservation 40(1):71–86.

Daily, G. C., S. Alexander, P. R. Ehrlich, L. Goulder, J. Lubchenco, P. A. Matson, H. A. Mooney, S. Postel, S. H. Schneider, D. Tilman, and G. M. Woodwell. 1997. Ecosystem services: Benefits supplied to human societies by natural ecosystems. Issues in Ecology 2 (Spring):1–16.

Davis, D. M., K. P. Reese, S. C. Gardner, and K. L. Bird. 2015. Genetic structure of greater sage-grouse (Centrocercus urophasianus) in a declining, peripheral population. Condor: Ornithological Applications 117:530–544.

de Valpine, P., and A. Hastings. 2002. Fitting population models incorporating process noise and observation error. Ecological Monographs 72:57–76.

DeWeerdt, S. 2002. What really is an evolutionary significant unit? Conservation Biology in Practice 3:10–17.

DOI [Department of Interior] and DOC [Department of Commerce]. 1996. Policy regarding the recognition of distinct vertebrate population segments under the Endangered Species Act. Federal Register 61(26):4722–4725.

Dunk, J. R., B. Woodbridge, N. Schumaker, E. M. Glenn, B. White, D. W. LaPlante, R. G. Anthony, R. J. Davis, K. M. Dugger, K. Halupka, P. Henson, B. G. Marcot, M. Merola-Zwartjes, B. R. Noon, M. G. Raphael, J. Caicco, D. L. Hansen, M. J. Mazurek, and J. Thrailkill. 2019. Conservation planning for species recovery under the Endangered Species Act: A case study with the northern spotted owl. PLoS ONE 14(1):e0210643.

Eckrich, C. A., M. J. Warren, D. A. Clark, P. J. Milburn, S. J. Torland, and T. L. Hiller. 2018. Space use and cover

selection of kit foxes (Vulpes macrotis) at their distributional periphery. American Midland Naturalist 179(2):247–260.

Engstrom, R. T., R. L. Crawford, and W. W. Baker. 1984. Breeding bird populations in relation to changing forest structure following fire exclusion: A 15-year study. Wilson Bulletin 96:437–450.

Fattebert, J., H. S. Robinson, G. Balme, R. Slotow, and L. Hunter. 2015. Structural habitat predicts functional dispersal habitat of a large carnivore: How leopards change spots. Ecological Applications 25:1911–1921.

Frankel, O. H., and M. E. Soulé. 1981. Conservation and evolution. Cambridge University Press, Cambridge.

Franklin, J. F. 1993. Preserving biodiversity: Species, ecosystems, or landscapes? Ecological Applications 3:202–205.

Furlow, F. B., and T. Armijo-Prewitt. 1995. Peripheral populations and range collapse. Conservation Biology 9:1345.

Galetti, M., R. Guevara, M. C. Côrtes, R. Fadini, S. Von Matter, A. B. Leite, F. Labecca, T. Ribeiro, C. S. Carvalho, R. G. Collevatti, M. M. Pires, P. R. Guimarães Jr., P. H. Brancalion, M. C. Ribeiro, and P. Jordano. 2013. Functional extinction of birds drives rapid evolutionary changes in seed size. Science 240(6136):1086–1090.

Gehrt, S. D., and J. E. Chelsvig. 2003. Bat activity in an urban landscape: Patterns at the landscape and microhabitat scale. Ecological Applications 13:939–950.

Graham, R. L., C. T. Hunsaker, R. V. O'Neill, and B. L. Jackson. 1991. Ecological risk assessment at the regional scale. Ecological Applications 1:196–206.

Gray, A. N. 2000. Adaptive ecosystem management in the Pacific Northwest: A case study from coastal Oregon. Conservation Ecology 4(2):6, https://www.ecologyand society.org/vol4/iss2/art6/.

Green, D. M. 2005. Designatable units for status assessment of endangered species. Conservation Biology 19:1813–1820.

Greenberg, R., and J. E. Maldonado. 2006. Diversity and endemism in tidal-marsh vertebrates. Studies in Avian Biology 32:32–53.

Grumbine, R. E. 1994. What is ecosystem management? Conservation Biology 8:27–38.

Gunderson, L. 1999. Resilience, flexibility and adaptive management—antidotes for spurious certitude? Conservation Ecology 3:7, https://www.ecologyand society.org/vol3/iss1/art7/.

Gunderson, L. H., C. S. Holling, L. Pritchard Jr., and G. D. Peterson. 2002. Resilience of large-scale resource systems. Pp. 3–20 in L. H. Gunderson and L. Pritchard Jr., eds. Resilience and the behavior of large-scale systems. Island Press, Washington, DC.

Guthery, F. S. 2008. A primer on natural resource science. Texas A&M University Press, College Station.

Haig, S. M., E. A. Beever, S. M. Chambers, H. M. Draheim, B. D. Dugger, S. Dunham, E. Elliott-Smith, J. B. Fontaine, D. C. Kesler, B. J. Knaus, I. F. Lopes, P. Loschl, T. D. Mullins, and L. M. Sheffield. 2006. Taxonomic considerations in listing subspecies under the US Endangered Species Act. Conservation Biology 20:1584–1594.

Hallfors, M. H., J. Liao, J. Dzurisin, R. Grundel, M. Hyvarinen, K. Towle, G. C. Wu, and J. J. Hellmann. 2016. Addressing potential local adaptation in species distribution models: Implications for conservation under climate change. Ecological Applications 26:1154–1169.

Harris, L. D. 1988. Landscape linkages: The dispersal corridor approach to wildlife conservation. Transactions of the North American Wildlife and Natural Resources Conference 53:595–607.

Harrison, S., and R. Noss. 2017. Endemism hotspots are linked to stable climatic refugia. Annals of Botany 119(2):207–214, doi:10.1093/aob/mcw248.

Haufler, J. B., R. K. Baydack, H. Campa III, B. J. Kernohan, C. Miller, L. J. O'Neil, and L. Waits. 2002. Performance measures for ecosystem management and ecological sustainability. Technical Review 02-1. The Wildlife Society, Bethesda, MD.

Havens, K. E., and N. G. Aumen. 2000. Hypothesis-driven experimental research is necessary for natural resource management. Environmental Management 25:1–7.

Henle, K., C. Andres, D. Bernhard, A. Grimm, P. Stoev, N. Tzankov, and M. Schlegel. 2017. Are species genetically more sensitive to habitat fragmentation on the periphery of their range compared to the core? A case study on the sand lizard (*Lacerta agilis*). Landscape Ecology 32:131–145.

Herman, S. G. 2002. Wildlife biology and natural history: Time for a reunion. Journal of Wildlife Management 66:933–946.

Hitchings, S. P., and T. J. C. Beebee. 1998. Loss of genetic diversity and fitness in common toad (*Bufo bufo*) populations isolated by inimical habitat. Journal of Environmental Biology 11:269–283.

Hof, J., and M. Bevers. 1999. Spatial optimization for managed ecosystems. Forest Science 45:595.

Holling, C. S. 1978. Adaptive environmental assessment and management. John Wiley & Sons, New York.

Hudson, W. E., ed. 1991. Landscape linkages and biodiversity. Island Press, Washington, DC.

Ibarguchi, G. 2014. From Southern Cone arid lands, across Atacama, to the Altiplano: Biodiversity and conservation at the ends of the world. Biodiversity 15:255–264.

Irwin, L. L., and T. B. Wigley. 1993. Toward an experimental basis for protecting forest wildlife. Ecological Applications 3:213–217.

James, S. E., and R. T. Coskey. 2002. Patterns of microhabitat use in a sympatric lizard assemblage. Canadian Journal of Zoology 80:2226–2234.

Jennelle, C. 2000. Avian communities, habitat associations, and reproductive success in even-age managed areas of Ouachita National Forest. Master's thesis. University of Arkansas, Fayetteville.

Johnston, D. W., and E. P. Odum. 1956. Breeding bird populations in relation to plant succession on the Piedmont of Georgia. Ecology 37:50–62.

Jorgenson, M. T., B. G. Marcot, D. K. Swanson, J. C. Jorgenson, and A. R. DeGange. 2015. Projected changes in diverse ecosystems from climate warming and biophysical drivers in northwest Alaska. Climatic Change 130:131–144.

Karl, S. A., and B. W. Bowen. 1999. Evolutionary significant units versus geopolitical taxonomy: Molecular systematics of an endangered sea turtle (genus *Chelonia*). Conservation Biology 13:990–999.

Keddy, P. A., H. T. Lee, and I. C. Wisheu. 1993. Choosing indicators of ecosystem integrity: Wetlands as a model ecosystem. Pp. 61–79 *in* S. Woodley, J. Kay, and G. Francis, eds. Ecological integrity and the management of ecosystems. St. Lucie Press, Delray Beach, FL.

Kerr, J. T., and D. J. Currie. 1995. Effects of human activity on global extinction risk. Conservation Biology 9:1528–1538.

Knapp, A. K., J. M. Blair, J. M. Briggs, S. L. Collins, D. C. Hartnett, L. C. Johnson, and E. G. Towne. 1999. The keystone role of bison in North American tallgrass prairie. BioScience 49:39–50.

Kolasa, J. 1989. Ecological systems in hierarchical perspective: Breaks in community structure and other consequences. Ecology 70:36–47.

Kvasnes, M. A. J., H. C. Pederson, T. Storras, and E. B. Nilsen. 2017. Vegetation type and demography of low density willow ptarmigan populations. Journal of Wildlife Management 81:174–181.

Lambert, B. A., R. A. Schorr, S. C. Schneider, and E. Muths. 2016. Influence of demography and environment on persistence of toad populations. Journal of Wildlife Management 80:1256–1266.

Lance, G. N., and W. T. Williams. 1967. A general theory of classificatory sorting strategies: I. Hierarchical systems. Computing Journal 9:373–380.

Lavoie, M., P. Blanchette, S. Larivière, and J.-P. Tremblay. 2017. Winter and summer weather modulate the demography of wild turkeys at the northern edge of the species distribution. Population Ecology 59:239–249.

Lawton, J. H., and V. K. Brown. 1994. Redundancy in ecosystems. Pp. 255–270 *in* E. D. Schulze and H. A.

Mooney, eds. Biodiversity and ecosystem function. Springer-Verlag, Berlin.

Lee, K. N., and J. Lawrence. 1986. Adaptive management: Learning from the Columbia River Basin Fish and Wildlife Program. Environmental Law 16:431–460.

Lesica, P., and F. W. Allendorf. 1995. When are peripheral populations valuable for conservation? Conservation Biology 9:753–760.

Lindenmayer, D. B., R. B. Cunningham, M. L. Pope, and C. F. Donnelly. 1999. A large-scale "experiment" to examine the effects of landscape context and habitat fragmentation on mammals. Biological Conservation 88:387–403.

Louda, S. M., R. W. Pemberton, M. T. Johnson, and P. A. Follett. 2003. Nontarget effects—the Achilles' heel of biological control? Retrospective analyses to reduce risk associated with biocontrol introductions. Annual Review of Entomology 48:365–396.

Lynam, T., F. Bousquet, C. Le Page, P. d'Aquino, O. Barreteau, F. Chinembiri, and B. Mombeshora. 2002. Adapting science to adaptive managers: Spidergrams, belief models, and multi-agent systems modeling. Conservation Ecology 5(2):24, https://www.ecologyand society.org/vol5/iss2/art24/.

Maes, J., B. Egoh, L. Welleman, C. Liquette, P. Vihervaara, J. P. Schager, B. Grizeiit, E. G. Drakou, A. LaNotte, G. Zulian, F. Bouraoui, M. L. Paracchini, L. Braat, and G. Bidoglio. 2012. Mapping ecosystem services for policy support and decision making in the European Union. Ecosystem Services 1:31–39.

Mallet, J., and M. Joron. 1999. Evolution of diversity in warning color and mimicry: Polymorphisms, shifting balance, and speciation. Annual Review of Ecology and Systematics 30:201–234.

Marcot, B. G. 2002. An ecological functional basis for managing decaying wood for wildlife. Pp. 895–910 in W. F. Laudenslayer Jr., P. J. Shea, B. E. Valentine, C. P. Weatherspoon, and T. E. Lisle, eds. Proceedings of the Symposium on the Ecology and Management of Dead Wood in Western Forests. General Technical Report PSW-GTR-181. US Department of Agriculture, Forest Service, Pacific Southwest Research Station, Albany, CA.

Marcot, B. G., and K. B. Aubry. 2003. The functional diversity of mammals in coniferous forests of western North America. Pp. 631–664 in C. J. Zabel and R. G. Anthony, eds. Mammal community dynamics: Management and conservation in the coniferous forests of western North America. Cambridge University Press, Cambridge.

Marcot, B. G., M. A. Castellano, J. A. Christy, L. K. Croft, J. F. Lehmkuhl, R. H. Naney, K. Nelson, C. G. Niwa, R. E. Rosentreter, R. E. Sandquist, B. C. Wales, and E. Zieroth. 1997. Terrestrial ecology assessment. Pp. 1497–1713 in T. M. Quigley and S. J. Arbelbide, eds. An assessment of ecosystem components in the interior Columbia Basin and portions of the Klamath and Great Basins, vol. 3. General Technical Report PNW-GTR-405. US Department of Agriculture, Forest Service, Pacific Northwest Research Station, Portland, OR.

Marcot, B. G., L. K. Croft, J. F. Lehmkuhl, R. H. Naney, C. G. Niwa, W. R. Owen, and R. E. Sandquist. 1998. Macroecology, paleoecology, and ecological integrity of terrestrial species and communities of the interior Columbia River Basin and portions of the Klamath and Great Basins. General Technical Report PNW-GTR-410. US Department of Agriculture, Forest Service, Pacific Northwest Research Station, Portland, OR.

Marcot, B. G., and M. Vander Heyden. 2001. Key ecological functions of wildlife species. Pp. 168–186 in D. H. Johnson and T. A. O'Neil, eds. Wildlife-habitat relationships in Oregon and Washington. Oregon State University Press, Corvallis.

Marcot, B. G., B. C. Wales, and R. Demmer. 2003. Range maps of terrestrial species and natural areas in the interior Columbia River Basin and northern portions of the Klamath and Great Basins. General Technical Report PNW-GTR-583. US Department of Agriculture, Forest Service, Pacific Northwest Research Station, Portland, OR, https://www.fs.usda.gov/pnw/node/27992/.

Margalida, A., J. A. Sánchez-Zapata, G. Blanco, F. Hiraldo, and J. A. Donázar. 2014. Diclofenac approval as a threat to Spanish vultures. Conservation Biology 28:631–632.

Masters, R. E., R. L. Lochmiller, S. T. McMurry, and G. A. Bukenhofer. 1998. Small mammal response to pine-grassland restoration for red-cockaded woodpeckers. Wildlife Society Bulletin 28:148–158.

Masters, R. E., C. W. Wilson, D. S. Cram, G. A. Bukenhofer, and R. L. Lochmiller. 2002. The influence of ecosystem restoration for red-cockaded woodpeckers on breeding bird and small mammal communities. Pp. 73–90 in W. M. Ford, K. R. Russel, and C. E. Moorman, eds. The role of fire in nongame wildlife management and community restoration: Traditional uses and new directions. General Technical Report NE-288. US Department of Agriculture, Forest Service, Northeastern Research Station, Delaware, OH.

Matlack, R. S., D. W. Kaufman, and G. A. Kaufman. 2002. Influence of grazing by bison and cattle on deer mice in burned tallgrass prairie. American Midland Naturalist 146:361–368.

Maxwell, S. L., N. Butt, M. Maron, C. A. McAlpine, S. Chapman, A. Ullmann, D. B. Segan, and J. E. M. Watson. 2018. Conservation implications of ecological responses to extreme weather and climate events. Diversity and Distributions 25:613–625.

McConkey, K. R., and D. R. Drake. 2006. Flying foxes cease to function as seed dispersers long before they become rare. Ecology 87:271–276.

McElroy, D. M., J. A. Shoemaker, and M. E. Douglas. 1997. Discriminating Gila robusta and Gila cypha: Risk assessment and the Endangered Species Act. Ecological Applications 7:958–967.

McNay, R. S., R. E. Page, and A. Campbell. 1987. Application of expert-based decision models to promote integrated management of forests. Transactions of the North American Wildlife and Natural Resource Conference 52:82–91.

Meyers, J. M., and A. S. Johnson. 1978. Bird community associated with succession of loblolly and shortleaf pine forests. Pp. 50–65 in R. M. DeGraff, technical coordinator. Proceedings of the Workshop on Management of Southern Pine Forests for Nongame Birds. General Technical Report SE-14. US Department of Agriculture, Forest Service, Southeastern Forest Experiment Station, Asheville, NC.

Miller, J. C., P. C. Hammond, and D. N. R. Ross. 2003. Distribution and functional roles of rare and uncommon moths (Lepidoptera: Noctuidae: Plusiinae) across a coniferous forest landscape. Annals of the Entomological Society of America 96:847–855.

Mooney, H. A., J. H. Cushman, E. Medina, O. O. E. Sala, and E. D. Schulze. 1996. What we have learned about the ecosystem functioning of biodiversity. Pp. 475–484 in H. A. Mooney, J. H. Cushman, E. Medina, O. O. E. Sala, and E. D. Schulze, eds. Functional roles of biodiversity: A global perspective. Springer-Verlag, New York.

Morell, V. 1996. Starting species with third parties and sex wars. Science 273:1499–1502.

Moritz, C. 1994. Defining "evolutionarily significant units" for conservation. Trends in Ecology & Evolution 9:373–375.

Noon, B. R., K. S. McKelvey, and B. G. Dickson. 2009. Multispecies conservation planning on US federal lands. Pp. 51–84 in J. J. Millspaugh and F. R. Thompson Jr., eds. Models for planning wildlife conservation in large landscapes. Elsevier, New York.

Oman, S. D. 2000. Minimax hierarchical empirical Bayes estimation in multivariate regression. Journal of Multivariate Analysis 80:285–301.

Orians, G. H. 1993. Endangered at what level? Ecological Applications 3:206–208.

Ostfeld, R. S., and K. LoGiudice. 2003. Community disassembly, biodiversity loss, and the erosion of an ecosystem service. Ecology 84:1421–1427.

Otway, N. M., C. J. A. Bradshaw, and R. G. Harcourt. 2004. Estimating the rate of quasi-extinction of the Australian grey nurse shark (Carcharias taurus) population using deterministic age- and stage-classified models. Biological Conservation 119:341–350.

Palacios, M. G., A. M. Sparkman, and A. M. Bronikowski. 2011. Developmental plasticity of immune defence in two life-history ecotypes of the garter snake, Thamnophis elegans—a common-environment experiment. Journal of Animal Ecology 80:431–437.

Pedersen, Å. Ø., E. M. Soininen, S. Unander, M. H. Willebrand, and E. Fuglei. 2014. Experimental harvest reveals the importance of territoriality in limiting the breeding population of Svalbard rock ptarmigan. European Journal of Wildlife Research 60:201–212.

Pennock, D. S., and W. W. Dimmick. 1997. Critique of the evolutionary significant unit as a definition for "distinct population segments" under the US Endangered Species Act. Conservation Biology 11:611–619.

Peters, R. H. 1991. A critique for ecology. Cambridge University Press, Cambridge.

Peterson, D. L., and V. T. Parker. 1998. Ecological scale: Theory and applications. Columbia University Press, New York.

Policansky, D. 1993. Uncertainty, knowledge, and resource management. Ecological Applications 3:583–584.

Pontarp, M., J. Ripa, and P. Lundberg. 2015. The biogeography of adaptive radiations and the geographic overlap of sister species. American Naturalist 186:565–581.

Posadas, P., D. R. Miranda-Esquivel, and J. V. Crisci. 2001. Using phylogenetic diversity measures to set priorities in conservation: An example from southern South America. Conservation Biology 15:1325–1334.

Purushotham, C. B., and V. V. Robin. 2016. Sky island bird populations isolated by ancient genetic barriers are characterized by different song traits than those isolated by recent deforestation. Ecology and Evolution 60:7334–7343.

Real, L. A., and J. H. Brown. 1991. Foundations of ecology: Classic papers and commentaries. University of Chicago Press, Chicago.

Reckhow, K. H. 1990. Bayesian inference in non-replicated ecological studies. Ecology 71:2053–2059.

Rieman, B. E., D. C. Lee, R. F. Thurow, P. F. Hessburg, and J. R. Sedell. 2000. Toward an integrated classification of ecosystems: Defining opportunities for managing fish and forest health. Environmental Management 25:425–444.

Robbins, K. 2012. An ecosystem management primer: History, perceptions, and modern definition. Akron Law Publications, University of Akron School of Law, Akron,

OH, http://ideaexchange.uakron.edu/cgi/viewcontent
.cgi?article=1000&context=ua_law_publications/.

Robinson, Martyn. 1999. A field guide to frogs of Australia. New Holland, Sydney.

Rodhouse, T. J., E. A. Beever, L. K. Garrett, K. M. Irvine, M. R. Jeffress, M. Munts, and C. Ray. 2010. Distribution of American pikas in a low-elevation lava landscape: Conservation implications from the range periphery. Journal of Mammalogy 91:1287–1299.

Russell, K. R., D. C. Guynn Jr., and H. G. Hanlin. 2002. Importance of small isolated wetlands for herpetofaunal diversity in managed, young growth forests in the Coastal Plain of South Carolina. Forest Ecology and Management 163:43–59.

Sarr, D. A., A. Duff, E. C. Dinger, S. L. Shafer, M. Wing, N. E. Seavy, and J. D. Alexander. 2015. Comparing ecoregional classifications for natural areas management in the Klamath region, USA. Natural Areas Journal 35:360–377.

Sasaki, T., T. Furukawa, Y. Iwasaki, M. Seto, and A. S. Mori. 2015. Perspectives for ecosystem management based on ecosystem resilience and ecological thresholds against multiple and stochastic disturbances. Ecological Indicators 57:395–408.

Saunders, D. A., G. W. Arnold, A. A. Burbidge, and J. M. Hopkins. 1987. The role of remnants of native vegetation in nature conservation: Future directions. Pp. 387–392 in D. A. Saunders, G. W. Arnold, A. A. Burbidge, and J. M. Hopkins, eds. Nature conservation: The role of remnants of native vegetation. Surrey Beatty & Sons, Chipping Norton, New South Wales, Australia.

Scheu, S., N. Schlitt, A. V. Tiunov, J. E. Newington, and T. H. Jones. 2002. Effects of the presence and community composition of earthworms on microbial community functioning. Oecologia 133:254–260.

Scott, M. J., G. R. Bilyard, S. O. Link, C. A. Ulibarri, H. E. Westerdahl, P. F. Ricci, and H. E. Seely. 1998. Valuation of ecological resources and functions. Environmental Management 22:49–68.

Sellman, S., T. Säterberg, and B. Ebenman. 2016. Pattern of functional extinctions in ecological networks with a variety of interaction types. Theoretical Ecology 9(1):83–94.

Semlitsch, R. D., and J. R. Bodie. 1998. Are small, isolated wetlands expendable? Conservation Biology 12:1129–1133.

Severinghaus, L. L. 1992. Monitoring the population of the endangered Lanyu scops owl (Otus elegans botolensis). Pp. 790–802 in D. McCullough and R. H. Barrett, eds. Wildlife 2001: Populations. Elsevier Applied Science, London.

Sheth, S. N., and A. L. Angert. 2016. Artificial selection reveals high genetic variation in phenology at the trailing edge of a species range. American Naturalist 187:182–193.

Shipley, J. R., A. Contina, N. Batbayar, E. S. Bridge, A. Townsend Peterson, and J. F. Kelly. 2013. Niche conservatism and disjunct populations: A case study with painted buntings (Passerina ciris). Auk 130:476–486.

Sims, K. E., J. M. Alix-Garcia, E. Shapiro-Garza, L. R. Fine, V. C. Radeloff, G. Aronson, S. Castillo, C. Ramirez-Reyes, and P. Yanez-Pagans. 2014. Improving environmental and social targeting through adaptive management in Mexico's payments for hydrological services program. Conservation Biology 28:1151–1159.

Smith, T. B., and S. Skulason. 1996. Evolutionary significance of resource polymorphisms in fishes, amphibians, and birds. Annual Review of Ecology and Systematics 27:111–133.

Spencer, R., and M. B. Thompson. 2003. The significance of predation in nest site selection of turtles: An experimental consideration of macro- and microhabitat preferences. Oikos 102:592–600.

Steen, D. A., and K. Barrett. 2015. Should states in the USA value species at the edge of their geographic range? Journal of Wildlife Management 79(6):872–876.

Storfer, A. 1999. Gene flow and endangered species translocations: A topic revisited. Biological Conservation 87:173–180.

Stroud, J. T., and J. B. Losos. 2016. Ecological opportunity and adaptive radiation. Annual Review of Ecology, Evolution, and Systematics 47:507–532.

Takekawa, J. Y., and E. O. Garton. 1984. How much is an evening grosbeak worth? Journal of Forestry 82:426–428.

Tappe, P. A., R. E. Thill, J. J. Krystofik, and G. A. Heidt. 1994. Small mammal communities in mature pine-hardwood stands in the Ouachita Mountains. Pp. 74–81 in J. B. Baker, compiler. Proceedings of the Symposium on Ecosystem Management Research in the Ouachita Mountains: Pretreatment conditions and preliminary findings. General Technical Report SO-112. US Department of Agriculture, Forest Service, Southern Forest Experiment Station, New Orleans, LA.

Tappe, P. A., R. E. Thill, M. A. Melchiors, and T. B. Widley. 2004. Breeding bird communities on four watersheds under different forest management scenarios in the Ouachita Mountains, Arkansas. Pp. 154–163 in J. M. Guldin, technical compiler, Ouachita and Ozark Mountains Symposium: Ecosystem management research. General Technical Report BRS-74. US

Department of Agriculture, Forest Service, Southern Research Station, Asheville, NC.

Tracy, C. R., and P. F. Brussard. 1994. Letters to the editor: Preserving biodiversity; Species in landscapes. Ecological Applications 4:205–207.

Treitz, P., and P. Howarth. 2000. Integrating spectral, spatial, and terrain variables for forest ecosystem classification. Photogrammetric Engineering and Remote Sensing 66:305–318.

Urban, M. C. 2015. Accelerating extinction risk from climate change. Science 348(6234):571–573.

Von Bertalanffy, L. 1968. General systems theory. George Braziller, New York.

Vongraven, D., and A. Bisther. 2014. Prey switching by killer whales in the north-east Atlantic: Observational evidence and experimental insights. Journal of the Marine Biological Association of the United Kingdom 94:1357–1365.

Vucetich, J. A., M. P. Nelson, and J. T. Bruskotter. 2017. Conservation triage falls short because conservation is not like emergency medicine. Frontiers in Ecology and Evolution 5:1–6, doi:10.3389/fevo.2017.00045.

Walker, B., A. Kinzig, and J. Langridge. 1999. Plant attribute diversity, resilience, and ecosystem function: The nature and significance of dominant and minor species. Ecosystems 2:95–113.

Walters, C. 1986. Adaptive management of renewable resources. Macmillan, New York.

Walters, C., J. Korman, L. E. Stevens, and B. Gold. 2000. Ecosystem modeling for evaluation of adaptive management policies in the Grand Canyon. Conservation Ecology 4:1, https://consecol.org/vol4/iss2/art1/.

Waples, R. S. 1995. Evolutionarily significant units and the conservation of biological diversity under the Endangered Species Act. American Fisheries Society Symposium 17:8–27.

Warren, C. D., J. M. Peek, G. L. Servheen, and P. Zager. 1996. Habitat use and movements of two ecotypes of translocated caribou in Idaho and British Columbia. Conservation Biology 10(2):547–553.

Warshall, P. 1995. The biopolitics of the Mt. Graham red squirrel (*Tamiasciuris hudsonicus grahamensis*). Conservation Biology 8:977–988.

Wellemeyer, J. C., J. S. Perkin, J. D. Fore, and C. Boyd. 2018. Comparing assembly processes for multimetric indices of biotic integrity. Ecological Indicators 89:590–609.

Westgate, M. J., G. E. Likens, and D. B. Lindermayer. 2012. Adaptive management of biological systems: A review. Biological Conservation 158:128–139.

Wetzel, F. T., H. Beissmann, D. J. Penn, and W. Jetz. 2013. Vulnerability of terrestrial island vertebrates to projected sea-level rise. Global Change Biology 19(7):2058–2070, doi:10.1111/gcb.12185.

Wiedmann, B. P., and G. A. Sargeant. 2014. Ecotypic variation in recruitment of reintroduced bighorn sheep: Implications for translocation. Journal of Wildlife Management 78(3):394–401.

Wiens, J. A. 2016. Ecological challenges and conservation conundrums. John Wiley & Sons, New York.

Wiens, J. J. 1999. Polymorphism in systematics and comparative biology. Annual Review of Ecology and Systematics 30:327–362.

Williford, D., R. W. DeYoung, R. L. Honeycutt, L. A. Brennan, and F. Hernandez. 2016. Phylogeography of the bobwhite (*Colinua*) quails. Wildlife Monographs 193:1–49.

Wilson, C. W., R. E. Masters, and G. A. Bukenhofer. 1995. Breeding bird response to pine-grassland community restoration for red-cockaded woodpeckers. Journal of Wildlife Management 59:56–67.

Wittmer, H. U., B. N. McLellan, D. R. Seip, J. A. Young, T. A. Kinley, G. S. Watts, and D. Hamilton. 2005. Population dynamics of the endangered mountain ecotype of woodland caribou (*Rangifer tarandus caribou*) in British Columbia, Canada. Canadian Journal of Zoology 83:407–418.

Wüest, R. O., A. Antonelli, N. E. Zimmermann, and H. P. Linder. 2015. Available climate regimes drive niche diversification during range expansion. American Naturalist 185(5):640–652.

Yadamsuren, O., J. D. Murdoch, E. Purevee, M. Munkhbayar, A. Jargalsaikhan, Z. Purevjargal, M. Khorloo, and T. Khayankhyarvaa. 2018. Estimating occupancy and detectability of toad headed agamas at the periphery of their range in Mongolia. Journal of Herpetology 52:361–368.

Young, H. S., D. J. McCauley, M. Galetti, and R. Dirzo. 2016. Patterns, causes, and consequences of Anthropocene defaunation. Annual Review of Ecology, Evolution, and Systematics 47:333–358.

Zhao, L. Z., A. S. Colman, R. J. Irvine, S. R. Karlsen, G. Olack, and E. A. Hobbie. 2019. Isotope ecology detects fine-scale variation in Svalbard reindeer diet: Implications for monitoring herbivory in the changing Arctic. Polar Biology 42:793–805.

Zorn, P., W. Stephenson, and P. Grigoriev. 2001. An ecosystem management program and assessment process for Ontario national parks. Conservation Biology 15:353–362.

7 — Putting Concepts into Practice
Guidelines for Developing Study Plans

> Once you learn to read the land, I have no fear of what you will do to it, or with it. And I know many pleasant things it will do to you.
>
> Aldo Leopold, in Fladder and Callicott (1992:337)

Throughout this book we have made the case for refocusing our studies of animal ecology on biological populations and concomitantly tightening up the way we design and implement our research efforts. We have repeatedly pointed out that we need to become much more aware of how we use and apply terminology, because we must all speak a common and well-defined language if we are to actually translate our science into meaningful and lasting actions in the environment. Our goal for this chapter is to provide a clear and practical framework within which we can actually advance our understanding of the ecology of wild animals and, most importantly, substantially improve how we communicate and how we use our research to enhance the conservation of animal populations.

General Concepts
Management: Populations versus Habitats

In his review of habitat studies, Morrison (2012) concluded that no major change had occurred in the way we have approached studies of wildlife-habitat relationships for at least the past 4 decades, and perhaps even longer. Most advances have come in the form of improved technology and more-sophisticated statistical analyses. We are not criticizing such advances per se; they are welcome and necessary. It seems that reviews of most journal manuscripts, however, focus on matters of technology and statistics, and they largely brush under the rug concerns about sampling design and the relevance thereof to the biological population (Block 2012). Because most habitat studies collected additional examples of phenomena that were already well studied, Morrison (2012) concluded that what we have called wildlife-habitat relationships had become outmoded (see also Morrison et al. 2012).

In a typical habitat study, following the wont of journal editors, a convenient research area is selected, samples of the usual vegetation variables and other environmental parameters are collected (often far more than are needed), a series of statistical analyses are conducted, and the results are compared with studies done at different times and (usually) different locations, with publication then justified by extrapolating the findings to some unspecified larger area. Lastly, additional research needs are enumerated, plus a list of generally vague recommendations for management, sometimes extrapolating far beyond the scope of the study and the area examined. In most cases, the animals under consideration are only a part (and an unknown part, at that) of the biological population, and any inferences drawn from such a study generally apply only to the place where and time when the research was conducted (Hurlbert 1984). The investigators may have measured dozens of variables on thousands of plots, but the reality is that they have an effective sample size of a single set of conditions, and extrapolation beyond that location is not necessarily justified. Even studies that incorporate measures of reproductive success or body condition (i.e., habitat quality) usually fail to delineate the known segment of the population. Are we sampling from the middle or an edge of the range of the species? Are we sampling from only a small portion of the spatial or seasonal conditions that the species can occupy within a portion of their range? Is what we are doing even of relevance to the biology of the species? The problem is that our local-scale research might yield results that are quickly swamped by interactions happening outside of our study area's boundaries, as well as being modified by the presence, abundance, and behavior of multiple other species.

Thus we seem to have things reversed. For example, it seems logical to first determine the biological population to be considered, and then design a way to adequately sample the characteristics of interest from that population, such as features of the environment that are being used (i.e., habitat). As noted above, however, with most of our wildlife studies, we first designate a convenient sampling area, find some individuals of interest, measure things about them, and hope the results mean something for a broader region. We think it unlikely that any reader would trust the results of a medical study where researchers visited the nearest shopping mall, merely took blood pressure readings from people they encountered, and used the results to make conclusions on their health. We are not calling for all studies to do all things. That would actually be counterproductive, because it is always better to do a few things and do them well, rather than to do a lot of things, but do them poorly. Our point is that all studies should be focused on a valid and rigorous sampling scheme across a biological population. We strongly encourage habitat studies that in-

corporate population parameters (e.g., breeding outcomes, gender-specific survival), but we recognize that this will not be practical for many projects (e.g., master's-level research). While such studies are necessarily limited, there is no reason why they cannot be developed on a foundation of what is known and is likely to be occurring in the actual biological population under investigation. In other words, the context relative to the biological population needs to be laid out clearly. In this chapter, we provide specific guidelines on 1 way to approach setting the context for field studies.

Planning Studies: Temporal and Spatial Considerations

Guthery and Strickland (2015) reviewed appearances of the term "habitat" and summarized its primary uses to mean (1) where an organism lives, which includes biotic and abiotic requisites over time; (2) an arbitrary area of interest (habitat) that contains demographically or behaviorally unique subareas (habitats), which, in turn, encompass descriptively unique subareas (habitats); and (3) gibberish—that is, the meaning becomes ambiguous and indecipherable.

The authors then broke these meanings into 3 categories: classic, hierarchical, and irrational. The classic outlook included definitions with meanings similar to "a place to live" or "natural abode" (see Chapter 2). Classic habitat is an existential matter— it either exists or it does not. They argued that the classic version of habitat was logically a dead-end concept, except for descriptive natural-history science. The hierarchical outlook was an arbitrary area (level 1) with (demographically, behaviorally) meaningful subdivisions (level 2) that then were subdivided (level 3) and might be subdivided further (level 4). This hierarchical view of habitat represented sequential parsing, which is the study of an entity composed of parts that are, in turn, composed of parts, and so on. Lastly—and our favorite—the irrational outlook included any use of habitat in an indecipherable manner. Guthery and Strickland (2015) provided examples of the often tautological and multiple meanings applied to "habitat" throughout the scientific literature. They asked if the classic definition of habitat held sway over the hierarchical definition, or vice versa, as the appropriate model for research and management and then commented that both seemed to have a place. The authors concluded, in apparent exasperation, that "both will be with us forever." These comments highlight the plight that researchers delving into animal-habitat relationships find themselves stuck within.

The hierarchical concept of habitat selection has been promoted as a way to help us understand how an individual decides (using innate and learned behaviors) where to settle; we then measure this settlement to describe habitat. Areas of various sizes (frequently defined as circular in shape, as a basis for measuring characteristics such as the percentage of shrub cover or the height of herbaceous plants) are often constructed around the settlement location (e.g., foraging region, nest or den site), with increasingly coarser measurements of the environment, calculated by their distance from the settlement location. Some studies also attempt to provide a statistic for the presence or abundance of a few key competitors or predators that have some sort of relationship with the focal species of interest. There is nothing incorrect with proceeding in this manner, as the results give us a picture of where the animal was located at the time of measurement and some notion of the conditions surrounding it when it was at this location. What these studies lack, however, is much (if any) knowledge of how the focal species will respond to its surroundings as biotic and abiotic conditions change over time, a statement that holds even when the investigations are repeated for multiple seasons or years. This is because habitat descriptions per se do not capture anything about the various dynamic processes that interact directly and indirectly with the focal species or group of species. Such descriptions are static, rather than dynamic.

It should also be noted that nesting circular areas for sampling purposes is not actually hierarchal. Hierarchies are defined by structures where the lower levels are bounded and enclosed by higher levels. For example, forests contain trees, that contain leaves, that contain cells, that contain chloroplasts, that contain DNA, that contain base pairs, and so on. Nested circles do not describe a hierarchy. Rather, they represent discrete points along a scaling gradient. The same features (e.g., a proportion of a specific land-cover type) are generally measured within all of the nested circles and compared across the circles' radii. Indeed, the move to coarser gauges at larger scales is driven more by how the data are obtained (e.g., plot-level vegetation metrics at small scales and remotely sensed cover types at larger scales) than by obligate shifts in what is being measured.

In actual hierarchies, you cannot quantify the same structures at lower levels, because the attributes of higher hierarchal levels are hidden—that is, you can look down but you cannot look up (e.g., you can measure photosynthesis at the tree, leaf, or cellular level, but you cannot calculate a tree's geometry by studying its cells). Thus a selection analysis that starts with an appropriate (for the species being studied) regional scale, finds areas that are used in different proportions, and then repeats the selection analysis by limiting the selection universe to those areas identified at the regional scale is hierarchal, because the selection analysis at each smaller scale is bounded by the results of selection at the next larger scale.

A key problem with viewing concentric areas as hierarchies is that the circle sizes only have a weak and tangential relationship to the relevant biological scales to which the organism is responding. Such studies merely give us a description of habitat use at a point in time (or several points, at best), and at various arbitrary spatial scales. They do little, if anything, to inform us about how the species of interest is occupying and persisting at a meaningful biological level. Additionally, using a hierarchical approach to habitat description is not equivalent to a landscape approach. We provide suggestions on how to formally pursue hierarchical studies later on in this chapter.

As we have noted repeatedly in this book, we understand that defining a biological population is complicated, because of discontinuities in its distribution that are difficult to recognize (see Chapter 5 for solutions). We also often know little about individual animal movements across any spatial scale. Regardless, we must first understand how the animals are assorted into meaningful, interacting groups of individuals before we can gather a proper sample. That is, in this regard we cannot let perfect be the enemy of good. If we design our research to specifically target meaningful population groups, the results will be far more interpretable and transferrable than those associated with an arbitrary area—even if our population delineation is far from perfect. The demography and viability of populations have not been emphasized in most wildlife-habitat studies. Instead, abundance, density, or occupancy has been the focus. Investigations often seek calculations of abundance or density over measures of vital rates and population response, because the former do not require us to know anything about the underlying population structure. Also, they are easier to determine.

As the number of species in a location alters, so, too, do the ways in which they interact among a changing number of individuals within each of those species. Likewise, these relationships vary across spatial extents. Moreover, additional and subtle interactions occur, resulting in changes such as sex and age ratios. As has been explained in Chapter 5, Smallwood (2001) showed how enlarging the spatial extent of a sampling area increasingly captured functionally significant demographic units of different taxa, including both predators and prey. Although this example is a simple case, it clearly shows how the demographic relationships being expressed (e.g., density) vary across different sizes of sampling areas. Because population viability considers how long a population will persist at designated levels into a specified future time, we must know how a species is structured into populations if we are to appropriately design our viability investigations. A primary

reason why we study animals and where they live (habitat) is to provide information that can help ensure their persistence into the future, which includes animals that are used for human consumption and sport hunting.

The sites where we take samples along environmental gradients relative to geographic locations of the species of interest are important, as the choice of these places will give us different relationships between a parameter and the response of the species (i.e., positive, negative, or neutral). Three major messages emerge. First, averaging across the gradient without regard to the underlying population structure would result in a wide variance in parameter estimates and give a false impression of how the species is differentially responding to more-local environmental conditions across space. Second, the variation in how the species responds to environmental conditions is a critical consideration, especially if source-sink dynamics (Pulliam 1988) come into play. Averaging abundance or reproductive output across a broader landscape would provide misleading information on how a species is performing within a smaller area. For example, a study occurring just in a source or a sink area would not represent the overall status of the population. Third, we would not know how the individuals inhabiting our snapshot in space (study area) were being influenced by adjacent individuals, or how relationships would change over time as individuals move across our study area. This neighborhood effect (as described by Dunning et al. 1992) depends on the juxtaposition of patches on the landscape, relationships of the target species with other species in adjoining patches, and ways in which the various species move among those patches. Simply declaring that the "study population was closed for the duration of the study," without offering empirical evidence, does nothing to mitigate these real-world dynamics.

Just as the species composition of animals and the ecological roles they play in any defined space is, to varying degrees, constantly changing and thus affecting environmental conditions, the vegetation and physical characteristics of an area can also fluctuate. As important as it is to determine the habitat of an organism in its current realized niche, it is also critical to anticipate how that habitat will change as species' niches morph—that is, as organisms readjust and respond to shifts in resource availability. Not planning research that captures these temporal processes will lead to snapshot studies that have only limited transferability, either to other sites or to the future management of the selected locale.

By not accounting in any manner for spatial and temporal dynamics, investigations become simple observational documentations—perhaps quite precise ones—of habitat relationships at a specific time and place. If we do not capture the broader dynamics, we cannot accurately predict how habitat relationships will change, given future environmental conditions. Meta-analysis has been recommended as a means of piecing together knowledge across many different studies (Hunter and Schmidt 1990; Spake and Doncaster 2017), thus potentially circumventing some of the problems associated with such localized foci. With meta-analysis, you combine the results of examinations across different sampling intensities and geographic extents to increase confidence that a relationship can be generalized across space and over time (Pena 1997; Johnson 2002). If conducted in a planned manner, repeating investigations in these dimensions certainly has merit, in order to better test and corroborate the robustness of patterns and relationships observed from single studies. Regardless of planning, however, such efforts do not account for any of the issues we have raised concerning broader-scale dynamics. Below, we develop a more holistic approach to studying wildlife, with the goal of maintaining species viability.

Supplementing Habitat: The Niche Concept

Habitat is widely used as a focal point for animal ecology studies, because it is easy to see, as well as to quantify, its primary components—namely, vegetation and other obvious structures (e.g., rocks, water).

Understanding the distribution of animals is, however, intimately tied to the concept of niche. Grinnell (1917) introduced the term "niche" to explain the distribution of a single bird species, based on spatial considerations (e.g., reasons for a close association with a vegetation type), dietary dimensions, and constraints placed by the need to avoid predators. Thus the Grinnellian niche included both a species' positional (i.e., habitat) and functional roles in the community. Elton (1927) later described a niche as the status of an animal in the community, and he focused on trophic position and diet.

A key advance in the niche concept was distinguishing a fundamental niche from a realized niche. The former is basically the suite of resources (biotic and abiotic) a species could use in the absence of interference by other species. A realized niche, however, is a (usually) reduced resource-use space inhabited by a species in the presence and influence of other species or vectors (notably competitors, predators, disease, and other adverse biotic influences) and the environment. Other, more complicated views of a niche also arose, following the pioneering work by Grinnell (1917), Elton (1927), and others (e.g., Hutchinson 1957).

We need to recognize that studying habitat alone can provide only limited insight into the factors responsible for animal survival and fitness and, thus, population responses to changing environments. Mathewson and Morrison (2015) concluded that the proliferation of habitat terms has largely been the result of failure to think about a niche concept when studying how and why animals are distributed as they are. A habitat concept, by itself, usually cannot describe the underlying mechanisms that determine survival and fecundity. Consider the concentric approach, discussed above, to describe den or nest locations. Such sites integrate dynamic niche properties, including requirements for survival and successful reproduction. But the nature of these properties is largely hidden and weakly inferred from measured covariates. Research has shown that a number of environmental factors can restrict the survival and productivity of a species (either across its entire range or a portion thereof), and the influence of any single factor is not necessarily additive to the influence of any other factors. That is, usually only 1 factor is limiting in any particular location and time (Liebig's law of the minimum, first expressed by Justus Liebig in 1840), and it is unlikely that the same factor will always do so, because natural variation causes a shift in the quality and quantity of resources (Wiens 2016).

Focusing on habitat alone is problematic, because the environmental features we measure can stay the same, while the use of important resources by an animal within that habitat—and, therefore, the realized niche—can change. For example, consider a bird foraging on shrubs, where the shrubs ("habitat") might not alter in their physical dimensions or appearance, but the type and size of insect prey associated with those shrubs might indeed change. If we describe habitat only as the structural or floristic aspects of vegetation, such as these prey-inhabited shrubs, we will often fail to predict the health of an organism, because we did not recognize constraints on the exploitation of other resources (e.g., the type and size of the insect prey) that are the proximate limiting factors (Dennis et al. 2003).

Wildlife Habitat and Landscape Ecology

Landscape ecology focuses on how elements, or patches, in the environment are configured relative to one another in an overall mosaic, and, in turn, how such structures influence ecological patterns and processes (Wiens and Milne 1989). As we have discussed earlier in this book (see especially Chapters 1 and 3), in landscape ecology there is no single specific spatial scale (spatial extent) that defines a landscape. Rather, in landscape studies of animal ecology, a landscape should be viewed from the perspective of the animal under investigation. Studies of multiple species are more complicated and should involve several spatial scales, representing each species' different behavioral response to environmental

conditions. As was noted by Wiens and Milne (1989), a landscape that is heterogeneous from the perspective of a ground-dwelling insect may, from the viewpoint of a foraging ungulate, only contain 1 or 2 seemingly homogeneous patches.

When considering a landscape, we should think not only of the patch being used, but also of both the potential influence of adjoining patches and connectivity among patches as influencing use. The function and existence of a location (patch) as habitat is entirely conditional on the attributes of the landscape in which it is embedded (as well as on the specific behavioral attributes of the study's species of interest). Describing a location (patch) as habitat per se, in isolation from these other influences, is clearly problematic. Mitchell and Powell (2002) reviewed the concept of linking landscapes to the fitness of an organism, where habitat is mapped and modeled as a response surface, depicting how conditions contribute to that fitness. This approach does not rely on the tenuous assumptions that patches are both discrete and homogeneous. Instead, it incorporates concepts of heterogeneity, matrix areas, and corridors. Such a map represents a testable hypothesis that can be challenged with empirical data.

Turner (2005) analyzed the development of landscape ecology in North America, noting that concepts emerging from the literature now permeate ecological research. She reviewed definitions of landscape ecology, noting that they all shared an explicit focus on the importance of spatial heterogeneity for ecological processes. Of particular relevance to considerations of wild animals, she noted that the scale at which characteristics of the surrounding landscape influenced a local response had been demonstrated for a variety of taxa. Turner (2005) did mention habitat, but she used the term in 2 distinct contexts: in the sense of a land cover, such as habitat fragmentation, habitat abundance, and habitat configuration; and, apparently, in a species-specific sense when discussing habitat suitability and habitat use.

As we have noted above, approaching habitat analysis from a hierarchical perspective is not equivalent to conducting a landscape analysis. The typical hierarchical approach simply increases the area under study, in order to determine what conditions (i.e., habitat) prevail in areas of increasing size around the animal (or group of animals) being investigated. This hierarchy does place the specific location of an organism into a broader context, such as noting that a gopher species is in a meadow surrounded by a deciduous woodland, which is part of a conifer forest. All of these spatial areas can be described by means of any number of variables. This is useful information and can be employed, for example, to talk about how many similar meadows occur throughout the forest. Likewise, we can describe various features—and have done so thousands of times in the literature—such as the nest site, breeding territory, home range, and so forth across increasingly larger spatial extents for individual animals. This could be helpful in predicting the potential distribution of the species across space. But these analyses are static, not dynamic, and are not reflective of population dynamics in space and time. This is particularly important when considering the annual cycle of a species, during which their needs and distribution can change. For example, many migratory birds select a locale in which to breed, establish a territory, and nest. Once the nestlings fledge, the area used may expand beyond that initial territory as the adults mentor their young on where and how to forage. As fall approaches, birds migrate along pathways that may include different conditions than those found within their breeding grounds, and they ultimately settle in a wintering area, which can be quite different from where they bred or were reared.

A General Framework to Guide Study Design

In the sections that follow, we outline some practical ways for researchers to design efforts that have an overall similar framework and incorporate the concepts we have developed throughout this book.

If such a broad and general basis is adopted, then individual studies will be much more relevant to broader spatial and temporal extents than is currently the case. In other words, we need to take a major leap forward in the way we conduct our investigations, because the incremental steps of better-designed technology and statistics have not resulted in major advances in knowledge, which means that we are not providing the data or the context necessary for making substantial improvements in the way we actually manage animals in the environment. We are not proposing some standard way of doing research. That would simply stifle innovation, even if it was possible to accomplish it. Rather, we are proposing a general framework that uses terminology in a proper and consistent manner and would clearly present both what is known about the distribution of the selected biological population(s) and where the study fits in geographically.

Species Assembly: Establishing the Context

In this section we are not attempting to assemble species into any type of ecological grouping per se (e.g., community, guild), nor are we proposing that particular factors (e.g., competition, predation) are driving those groupings. Additionally, we are not recommending that any specific spatial scale be applied (as has been emphasized above, determining the appropriate spatial extent is the first priority for an investigation). Rather, we are borrowing from the ideas of species assembly rules as a way to set any examination of animal ecology into an overall context and help decide which biotic and abiotic factors should be addressed within the confines of the goals of a specific study.

Here we focus on identifying the filters and constraints that will modify the species present in an area throughout a successional pathway. Next, we summarize the basic assembly process, and then explain how it can be adapted to help the conceptualization of how we study species distribution and persistence.

Species Pool

Van Andel and Grootjans (2006) summarized the species pool concept, where a regional species pool occurs within a biogeographic region, thus encompassing multiple local species assemblages. For example, the number of species becomes increasingly smaller as we focus from a watershed extent down to a stream reach extent (Fig. 7.1). By passing through a series of filters, a local assemblage emerges from the larger pool. Van Andel and Grootjans (2006) defined the following types of pools; we retain their terminology but note that each step in the hierarchy is actually a subset of the larger pool of potentially occurring species.

- Regional species pool—the set of species occurring in a certain biogeographic or climatic region that are potential members of the target assemblage.
- Local species pool—the set of species occurring in a subunit of the biogeographic region, such as a valley segment.
- Community species pool—the set of species present in a site within the target community ("community" and "assemblage" would be synonymous).

The concept of ecological filters forms 1 of the main approaches in assembly rules theory (Hobbs

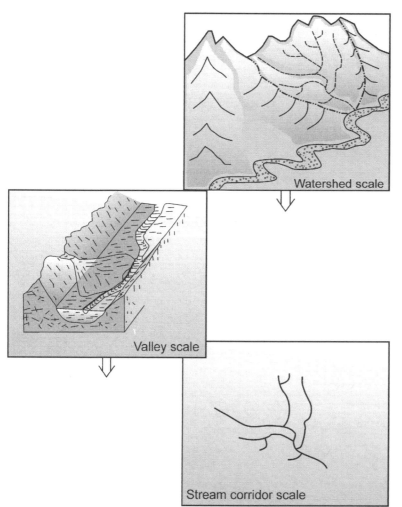

Figure 7.1 A schematic of the hierarchical scales used in watershed, riparian corridor, valley segment, and stream reaches. (Reproduced from Guisan and Thuiller (2005), with the permission of John Wiley & Sons)

and Norton 2004). Of the total pool of potential col-
onists, only those that are adapted to the abiotic and
biotic conditions present at a local site will be able
to become established. Thus a process of species
deletion or filtering occurs. Hobbs and Norton
(2004) applied assembly rules to restoration ecology, because their use at the beginning of restoration projects determined which factors were probably limiting the species present in the project area
in the first place. For example, Sundermann et al.
(2011) showed that of the 24 restoration projects
(treatment-control approach) and the assemblages
in the vicinity of the restored sites they reviewed, a
relationship could be found between the species
added at those sites and the regional species pool.

Likewise, we can apply the concept of species assembly to initially hypothesize the key parameters
(filers and constraints), spatial extent, and interactions of potential influence in the system we are
studying. It becomes our broad framework, which
incorporates habitat. This construct will apply regardless of the taxa of primary interest. Assembly
rules range from the obvious, such as that predators
without prey will starve, to the more complex, such
as the abundance and diversity of prey necessary for
a new predator to be present (Temperton and Hobbs
2004). It is also critical to recognize that the action
of filters will change over space and time.

Hobbs and Norton (2004), as modified by Morrison (2009), identified the following major filters.
Abiotic filters, with a few examples of each, include

- climate, such as rainfall and temperature
  gradients;
- substrate, such as fertility, soil water availability,
  toxicity; and
- landscape structure, such as landscape position,
  previous land use, patch size, and isolation.

There are also many biotic filters.

- Interspecific competition—from preexisting and
  potentially invading species, as well as between

planted or introduced species (intraspecific
competition, such as through density-dependent
factors, could also play a role).
- Predation-trophic interactions—from preexisting
  and potentially invading species, as well as from
  predation between reintroduced animal species.
- Propagule availability (dispersal)—bird perches,
  proximity to seed sources, presence of seed banks.
- Mutualisms—mycorrhizae, rhizobia, pollination
  and dispersal, defense, and the like.
- Natural or human-induced disturbance—
  presence of previous or new disturbance regimes.
- Order of species arrival and successional
  model—facilitation, inhibition, dispersal
  capabilities, and tolerance.
- Current and past composition and structure
  (biological legacy)—how much original biodiversity and original biotic and abiotic structure
  remains (the time frame for determining
  "original" must be stated and justified).

At least 7 of the figures in Temperton et al. (2004)
depicted the generalized pathway from a potential
pool of species, through various abiotic and biotic filters and other constraints, to the realized species in
a location. Morrison (2009:Fig. 5.2) synthesized
those figures into a single diagram depicting pathways and filters, showing how species fit into an
available niche space throughout the course of successional prey (Fig. 7.2).

As was shown for fish assemblages by Mouillot
et al. (2007), the latitudinal gradient in water temperature was the most influential factor explaining
richness distribution. At the regional scale, assemblage diversity was predictable, based on a limited
number of factors. At the local scale, fish richness
was determined by abiotic filters and biotic interactions acting simultaneously. As further reviewed by
Mouillot et al. (2007), the niche filtering hypothesis stated that coexisting species are more similar to
one another than would be expected by chance,
because of the filtering effect of environmental conditions that only allowed species with similar traits

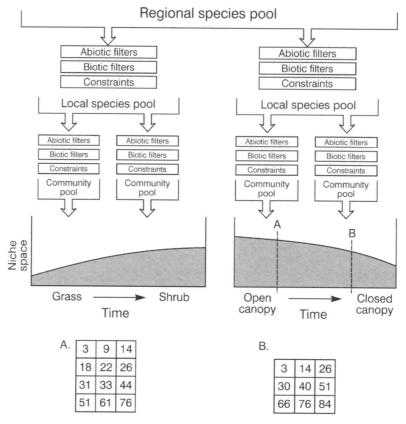

Figure 7.2   The species present at any area (community pool) are those remaining after the filtering process has occurred at the local and regional levels. The figure represents 2 local species pools drawn from the same regional pool that co-occur in time in 2 types of vegetation. The community pool associated with a seral stage is drawn from the local pool that is specific to the more general vegetation type. The type and number of species present across the seral stage will also be a reflection of the size of the target area and the niche space available. The cross-sectional cut depicted in the figure as (*A*) and (*B*) indicates how species (*numbered squares*) change in type and total number. (Adapted from Morrison (2009:Fig. 5.2))

to occur there. Additionally, biotic interactions further limited species presence at the local (pool) scale, because of factors such as competitive exclusion and limiting similarity principles.

Instructive examples of this process can be observed in the invasion of exotic species. These are introduced as new members of the regional species pool, and invasion occurs where allowed by abiotic and biotic filters. Those places where exotics are able to invade might contain species similar to the invader, since the native species have survived the same filtering process. As such, it is not infrequent that the exotic species can competitively

exclude native species and eliminate them from the system where both had been present. Again, fish provide a good example. Of the many species of exotic fish that have been introduced, the invasion of small cold streams is largely limited to salmonids (i.e., trout), due to abiotic filters. As most trout occupy very similar niches, it is common for exotic trout to displace the native ones. This has occurred in countless streams across the western United States, where eastern brook trout (*Salvelinus fontinalis*) have successfully colonized and subsequently excluded cutthroat trout (*Oncorhynchus clarkii*).

Time and Succession

The succession concept emerged from vegetation ecology, whereas the assembly rules concept arose from studies of animal communities in the context of island biogeography. The integration of paradigms such as these is critical in moving toward a better understanding of how organisms define their habitats across time and space (Pickett et al. 1994). Filters, assembly rules, disturbances, and succession are all intertwined. Studies of disturbances and succession contribute to our understanding of the processes that shape species distribution (Nuttle et al. 2004; White and Jentsch 2004).

As discussed above, the regional species pool of van Andel and Grootjans (2006) is filtered through various abiotic and biotic factors to result in the local species set. This process of filtering out species that are not adapted to a specific set of conditions must include a temporal component, because conditions change over time as succession proceeds. Thus we quickly see how this assembly approach incorporates time, which is missing, except in a static sense (e.g., a specific breeding season), from most habitat studies. In a successional landscape, plants, such as trees, differ from animals in important ways. Like animals, they are filtered from a larger pool, and that filtering process shifts over time. Unlike animals, however, individual trees of a given species can be highly variable in size and in their structural characteristics. As such, species composition and age are crucial elements in the biotic filtering. In studies of animal ecology, we often treat the structural attributes of a vegetation community as being static, but this is a function of the short duration of most animal habitat research efforts. Critically, even if the set of plant species is invariant, the structural characteristics they provide shift as they age.

Wiens (1989:74–77) explained that even if colonization of an area occurs, a species cannot be considered a bona fide member of a local community unless it is also established as a viable reproductive population. He further noted that because home-range size might not allow occupancy by enough individuals to develop a viable breeding population, a species might not permanently colonize the area. Figure 7.3 is a diagram of community assembly, reproduced from Wiens (1989:Fig. 4.1; see also Wiens 2016), which shows that the establishment of a population should proceed a researcher's designation of the community. He also stated that the activity of a predator could prevent colonization and the retention of a sufficient number of individuals within an area. Here we suggest that if the study area is smaller than the necessary size for population establishment of the most spatially demanding species in the regional pool, community understandings will be biased and incomplete.

Delineation of a study area must also take into account that the species mix is in flux, both seasonally and annually, as species leave or enter the community. Cam et al. (2002) noted that the species present in an area during a given year were a selection of those from what the authors termed a "supercommunity" (i.e., a pool). They assumed that the process, which resulted in the group of species that was present, involved local species extinctions and colonizations. Cam et al. (2002) suggested that such events were often stochastic and, hence, difficult to predict. The take-home point here is that sampling an area to delineate a community requires data from multiple years, to account for species moving into and out of the area. Further, if the study site was arbitrarily defined, it will be virtually impossible to extrapolate results to different times and, certainly, different places. Simply calling the animals in the spatial area that you have chosen to study a "community" is applying a fancy name to "that rather arbitrary gathering of animals I decided to research." We reiterate—knowing as much as you can about the biological populations of the species (or the area) under investigation is the essential first step in advancing what we call studies of community ecology.

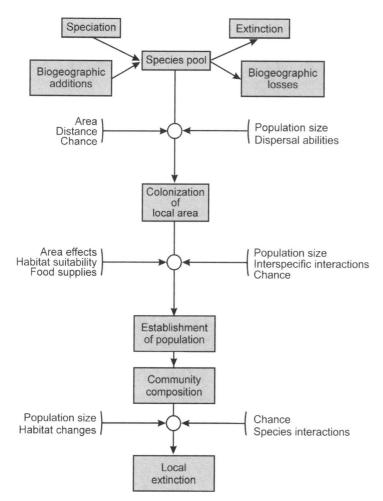

Figure 7.3   Factors contributing to the process of assembling communities of species. (Reproduced from Wiens (1989:Fig. 4.1), with the permission of Cambridge University Press)

Plan Implementation
Taking an Initial Step Forward

The assembly process discussed above directly influences each population segment through different combinations and strengths of filters and constraints. Thus the specific combination affecting, say, a particular subpopulation will probably be different in another subpopulation. These influences will change over short and long time frames. If we do not know the basic population structure, then we have little hope of knowing what is causing changes in species distribution, density, vital rates, and, ultimately, persistence. Nor will we know why local species pools

have the composition that they do. As we have suggested above, focusing on a small area, or averaging information across a broad area that is occupied by a species, usually results in no substantial gain in understanding changes in the distribution of that species or the composition of a local species pool.

There are many ways to gather data on the ecology of organisms and their environment, and there are various well-developed study designs that are appropriate, depending on each situation (e.g., Morrison et al. 2006, 2008). In Figure 7.4, we present a simplified version of the major approaches for studying animals. We acknowledge that there are often many plausible reasons given for the choice of a design.

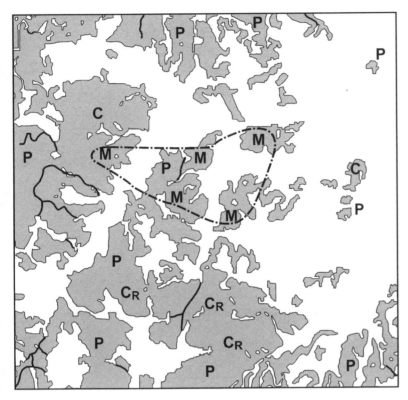

Figure 7.4 A depiction of the major approaches to obtaining data for studies of animal ecology. The classic design (*C*) is shown as a single study area, within which data on (usually) multiple individuals are gathered. C_R is the classic design with replication. *M* refers to a metapopulation approach, where the *dashed line* denotes a subjective boundary area. Lastly, *P* indicates a population-level approach that samples across the distribution of a biologically relevant entity.

Here we are focusing on the inherent limitations of various designs, relative to one based on the biological population.

A classic design involves the selection of a single study area, based on convenience and funding requirements. Study area size varies widely, depending on the objectives and available resources. Samples are gathered within that site, and investigators often claim that replication is obtained through the placement of points, transects, or subplots within the area Nonetheless, the term "pseudoreplication," accounting for potential autocorrelation among the samples, is more appropriate for such designs (Hurlbert 1984, 2009). What many would agree is an improvement over the classic design is the inclusion of true replications of study areas (i.e., classic with replication). Here, the selection of study area size, number of replicates, and juxtaposition of study areas will be based on convenience, logistics, and funding priorities.

Research on community ecology is usually conducted within these classic designs, with or without replication. The expansion of a community study to a network of communities represents the metacommunity approach. Our representation in Figure 7.4 is simplistic, and the dashed line is meant to indicate that the entire patch will be sampled in some manner (e.g., transects, points, subplots). The difference between this metacommunity approach and a classic one with replication (discussed above) is that in the former, investigators are also interested in exploring the dynamics among the patches and will often directly (empirical data) or indirectly (modeling) address various hypotheses (Leibold et al. 2004:Table 1). We suppose that a person could further replicate the metacommunity approach across a broader landscape, but we did not attempt to show that in this figure.

Lastly, Figure 7.4 depicts a rough approximation of how one would approach a population-based study.

Samples of patches are selected according to criteria that can be justified as relevant to the research question. For example, working with golden-cheeked warblers (*Setophaga chrysoparia*), Collier et al. (2012) used a stratified random sampling of patches of varying sizes, further stratified by their position within the breeding range of the species, to identify all patches of mature oak-juniper (*Quercus-Juniperus*) woodland known to be at least minimally acceptable to these warblers (see also Mathewson et al. 2012). They then used an analysis of sample size to guide their field sampling intensity (i.e., number of samples to collect). A similar approach was used to study black-capped vireos (*Vireo atricapilla*) across their Texas and Oklahoma breeding range (McFarland et al. 2013, 2015). By focusing on the overall population structure, along with consideration of previous research, these 2 bird species were shown to not be metapopulations. For the warblers, that research did, however, identify substantially different patterns of density, descriptions of habitat, and song structure, based on their geographic location within the species' range. We should mention that the studies exemplified here, as well as our discussion of replication, pertain largely to the use of frequentist statistical methods and their assumptions regarding autocorrelation, collinearity, and other considerations. Different research designs—such as before-and-after control-impact (BACI) designs and modifications thereof (e.g., Benedetti-Cecchi 2001), space-for-time replication (e.g., Covert-Bratland et al. 2006), or the use of Bayesian statistical methods (e.g., Duncan and Vesk 2013), might serve to modify or relax some of our suggested framework standards. Below, we outline steps to follow in the application of our framework, with a goal of substantially advancing our knowledge of how animals are distributed in space and time, and the reasons underlying their persistence.

Step 1. *Using the best information available, summarize the population structure of the target species (or multiple species).* This summary could be descriptive and qualitative, including the use of maps and remote sensing imagery. The goal here is to obtain the best

understanding possible of the range and distribution of the species, including the potential for a metapopulation or some other population measure, such as isolation by distance; effective population size (number of breeding individuals in a defined area); census population size (density or abundance of the species over some defined area); or genetic neighborhood (an area defined by where the parents of an individual could be from). Also include the designation of recognized or suspected subspecies, ecotypes, and isolated segments. This information provides an initial basis for study design. It also allows you to identify these details in subsequent reports and publications, so the context of your work is transparent and, as such, will serve as a more advanced starting point for future research by you or another investigator.

At a bare minimum, always discuss what is known about the population(s) under study, including potential subspecies, the possibility of ecotypes, and position of the species in the overall distributional range (e.g., center or edge), along with a consideration of breeding and social systems. Additional information on the presence-absence (better yet, density and activity) of possible predators, competitors, diseases and disease vectors, and other potentially interacting organisms further sets the context for a specific study, which will enable other investigators to better compare their work with your published information. Subsequent steps will be based primarily on the study's goals. We cannot anticipate all potential goals in our framework. Hence, below we provide generic guidance, including examples.

Step 2. *Design a study that obtains samples from across a relevant portion of the biological distribution of the target species, rather than simply selecting an area of convenience.* If funding or other constraints dictate the research area, at a minimum this procedure means that you are aware of where your study animals reside, relative to their overall range—that is, on an edge or near the center (sensu Van Horne 2002; see Step 1 above). Potentially, you will be able to sample across biologically relevant portions of the population. Even being restricted to a subset of known or probable disjunct segments provides a rigorous and justifiable design. For example, quantifying density and vital rates within several segments, and looking for dispersal between them, would provide valuable data if you are investigating the presence of a metapopulation. You will also be gathering data that is directly relevant to habitat quality.

Previously in this chapter, we have provided examples of studies that have successfully delineated biological populations, so specific guidance exists and is part of a growing body of literature. Moreover, if it is feasible, incorporate at least a cursory exami-

nation of the genetic makeup of individuals from across the geographic area you have identified as likely to be biologically relevant. This will provide substantial insight into the validity of your initial assumptions about population structure, as noted above. For example, although Anderson et al. (2018) concluded that they had identified a biological population of Townsend's big-eared bats (*Corynorhinus townsendii*), uncertainly remained as to the geographic extent and boundary of the population to the south and east of the initial sampling area (Fig. 7.5). Thus, as a next step in further delineating

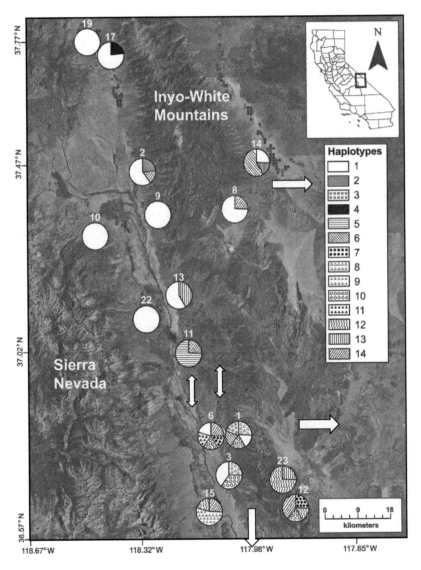

Figure 7.5 A haplotype network of Townsend's big-eared bats (*Corynorhinus townsendii*) in eastern California, mapped over the sampling region. (Updated from Anderson et al. (2018:Fig. 5), based on additional data)

the population, the researchers have initiated additional intensive sampling toward the east, as well as broader sampling across California and adjacent Nevada (M. L. Morrison, pers. obs.).

Step 3. *To further establish the context of your study for use in achieving additional goals (see below), or to provide a thorough description of the biotic and abiotic factors operating during your study (i.e., for journal papers and future research), determine the animal species that could potentially influence the behavior of the target species (or group of species).* This is part of the regional species pool. It will often include species that occupy much larger areas than those used by the target species, but it is a needed step in developing a valid landscape context. This initial list will be reduced in a subsequent step.

Step 4. *Determine additional potential filters and constraints that could modify the behavior, abundance, movements, and other aspects of the biology of species in the regional species pool.* Much of this work will involve literature reviews, along with expert opinion as needed. Remember, other species that may affect the study-specific target species could include organisms from a wide array of other major taxonomic groups. Examples include species of truffles eaten by flying squirrels; mammalian predators on avian species; domestic livestock as disease vectors to wild ungulates; and fossorial burrowing mammals, such as badgers, that provide habitat for burrowing owls (*Athene cunicularia*) and black-footed ferrets (*Mustela nigripes*).

Step 5. *Given the current condition of the sampling region, apply filters and constraints and determine the resulting local species pool.* This process could involve several likely scenarios, based on your knowledge of the study's particular environment. You have now developed a list of species that are likely to occur across the biologically relevant area under observation, as well as an understanding of those factors in need of quantification. This process should be adjusted to apply to each known or suspected population segment being examined.

Step 6. *If you are interested in projecting how the distribution and abundance of your target species could change over time, then you will need to determine the likely scenarios for natural disturbances, climate change, management actions, and plant succession over a reasonable and relevant time frame, such as 10 demographic generations of the target species.* This could include various outcomes, based on planned management activities (e.g., burning, clearing). Modeling changes in vegetation and other ambient conditions over time—such as with species distribution models (see Box 7.1)—would be a highly useful procedure. This should include a subsequent reevaluation of the regional species pool, filters, and constraints that might be related to future conditions (e.g., based on intervals of 10 years, or every few generations). This step is critical for forecasting changes in the landscape configuration and concomitant variations in animal distribution and abundance, because it includes related shifts in other species that influence the target species (or, if the animal assemblage is your focus, how it changes over time).

At a minimum, the above general procedure should result in predicting how the target species (or assemblage of species), as well as the larger pools of species, will vary in their distribution and abundance (given the methods used) as the environment alters over time. You will also generate what we currently would consider a typical description of habitat use, but within a much more biologically relevant spatial extent, and with knowledge of why the changes in use are occurring (e.g., succession, plant species diversity, predator-competitor occurrence). Because this procedure is based on a spatial extent that is relevant to the biological population and to other species influencing the target species, you will also have produced a more realistic representation of the landscape of the target species. Moreover, you will have followed a process that is clearly defined and justifiable at each step. An additional advantage is that someone—you or another researcher—could modify your input (e.g., species pools, filters, constraints) and explore corrected or alternative scenarios as new information is obtained.

Box 7.1. Species Distribution Models

There is a wide variety of analyses available that generally fall under species distribution models (SDMs). As thoroughly reviewed by Guisan and Thuiller (2005) and Elith and Leathwick (2009), these models relate field observations (e.g., presence-absence, abundance) to environmental predictor variables that are based on statistically or theoretically derived response. Environmental predictors are chosen from the 3 main types of influences on the species—namely, limiting factors, disturbances affecting the environmental systems (natural or human-induced), and resources. A core characteristic of SDMs is their relationship to the niche concept. Thus SDMs are directly practicable as tools for studying and developing management applications in applied ecology and wildlife biology.

SDM models can be constructed so the relationships between species and their overall environment are evident at different spatial scales. As had been noted by Guisan and Thuiller (2005:Fig. 1), the distribution observed over a large spatial extent and at coarse resolution is likely to be controlled by climatic regulators, whereas a patchy spatial distribution observed over a smaller area and at fine resolution most likely is related to a patchy

distribution of resources caused by natural topographic variations or human-induced fragmentation (Box Fig. 7.1a). Among the many applications provided by Guisan and Thuiller (2005:Table 1) were studies that (1) quantified the environmental niche of species; (2) tested biogeographical, ecological, and evolutionary hypotheses; (3) assessed species invasion and proliferation; (4) evaluated the impact of climate, land use, and other environmental changes on species distributions; (5) supported appropriate management plans for species recovery; (6) mapped suitable sites for species reintroduction; (7) supported conservation planning and reserve selection; and (8) modeled species assemblages (biodiversity, composition) from individual species predictions.

An output of SDMs are habitat maps, which have a clear and direct utility in describing animal habitat use, as well as changes in that use through space and time, and lay the foundation for analyzing the impact of potential natural and human-caused environmental changes in the future. One such example is from Johnson and Gillingham (2005:Fig. 2) Box Figure 7.1b provides an example of a species distribution map for woodland caribou (*Rangifer tarandus caribou*), based on ecological niche models using variables representing the

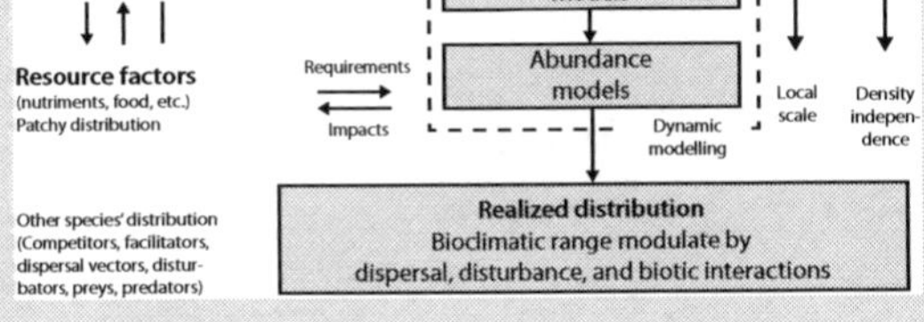

Figure 7.1a A general hierarchical modeling framework that depicts a means of integrating disturbance, dispersal, and population dynamics within species distribution models. (Reproduced from Guisan and Thuiller (2005:Fig. 1), with the permission of John Wiley & Sons)

distance of caribou locations from the nearest patch, the density of land-cover patches, or both patch density and distance. There are numerous statistical packages directly related to SDM analysis that use a wide variety of statistical methods, including artificial neural networks, Bayesian approaches, classification and regression trees, ecological niche factor analyses, both generalized additive and linear models, and other techniques (see Guisan and Thuiller 2005:Table 3).

As noted by Guisan and Thuiller (2005), a lack of knowledge about the focal species and the assemblage of species substantially complicates the development of a study of species distributions. Of particular relevance here is their call for information on home-range sizes and how the species of interest uses resources in the landscape. They noted that the choice of a geographic extent probably depends on existing knowledge of environmental gradients in the study area and of how different ages and sexes of the animals use these resources across seasons. The concept and application of species assembly we developed above provides the framework necessary to fulfill the requirements outlined by Guisan and Thuiller (2005). They specifically called for the collection of preliminary field observations, the application of sensitivity analyses, and the execution of experiments that quantify the fundamental range of the organisms' tolerance of predictor variables. Stepping through the development of species assembly provides an overall framework for any study, identifies both the data needed and the species to which it applies (e.g., as gathered through existing information, expert opinion, preliminary studies), and specifies the appropriate spatial scales.

Applications of SDMs have emphasized plants and fish, with fewer for wildlife species.

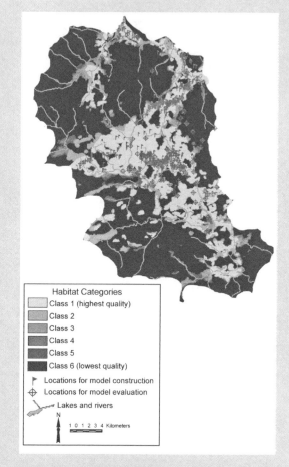

Figure 7.1b   Example of a species distribution map for woodland caribou (*Rangifer tarandus caribou*), based on ecological niche models using variables representing the distance of caribou locations from the nearest patch, the density of land-cover patches, or both patch density and distance. Species distribution maps were generated at a cell resolution of 25 × 25 m. A 6-class ranking scheme was used, where it was assumed that class 1 habitats were of the highest quality and class 6 were of the lowest. A validation set of caribou locations was overlain on the map and the number of locations within each of the 6 habitat classes was normalized by the area determined for that class. (Adapted from Johnson and Gillingham (2005:Fig. 2), courtesy of C. J. Johnson)

Many SDMs have used combinations of focal species distributions, measures of vegetation and topography, and climatic data in the development of models (Leyt et al. 2013;

Pikesly et al. 2013; Shirley et al. 2013; Bucklin et al. 2015; Young and Carr 2015). Pikesly et al (2013) used telemetry data from olive Ridley turtles (*Lepidochelys olivacea*) and information on their physical and biological environments to develop an SDM for this species. Lyet et al. (2013) combined species location data with climate and land-cover data to develop an SDM for the endangered Orsini's viper (*Vipera ursinii*), and Shirley et al. (2013) evaluated various remote sensing tools to models distributions of 40 land birds in western Oregon. Missing from many of the models was the inclusion of other species and how they might have influenced the distribution of the focal species. We envision this as the next logical step in improving the predictive capabilities of SDMs. Thus SDMs appear to offer a path in advancing our understanding of how animals are distributed in time and space.

We are not suggesting a fundamentally new theory or analytical procedure. Rather, we are strongly suggesting that animal ecologists need to move forward from concepts such as communities and metacommunities and reorient their efforts toward a biological population as the fundamental basis for study, within the context of how other organisms could affect that population.

Plan Implementation Steps: Managing the Environment

One of our goals for this book is to provide specific guidance on how to advance the study of animal ecology. Chapter 1 has argued that all of us, as wildlife specialists, have been stuck in a rut, so to speak, just repeating the same type of work on different species in various places and times, with no real recognition of how our results related to the actual biological population of interest. In Chapter 5, we have described specific ways to delineate biological populations. In this chapter, we have outlined ways to view and address management of the environment, with a goal of conserving biological populations. We have especially drawn on various studies and recommendations from the allied field of ecological restoration, as applied to animals (Morrison 2009). Although we are not trying to be prescriptive, we want to give enough detail and guidance so researchers can use our ideas as a starting point for more-specific plans.

Morrison (2009) synthesized the literature on ecological restoration and presented steps for wildlife restoration planning that were structured in a spatially explicit context, starting from a broad extent and moving down to local-scale applications. Below we generally follow his outline, but we have modified it to better reflect this book's focus on biological population(s) and an animal-centric (rather than human-centric) perception of the environment. Each step in your study's selection of species should be clearly documented in all relevant planning processes. We use the term "restoration" in a broad context and will not debate herein what the goals for restoration might be, as these vary, given the perspectives of the people doing the planning. The following is our 3-step framework for effective planning in the analysis and management of populations: defining the planning area, delineating the project area, and establishing an adaptive management plan.

Step 1. *Define the planning area (e.g., a basin composed of multiple watersheds).* When discussing a planning area (geographic extent) for any study—and, more specifically, for a management action (such as restoration)—Morrison (2009) took the perspective of starting spatially large and then working sequentially down to the needs of individual animal species within the planning area (a temporal component would also apply across all spatial scales). "Large," however, will be dependent on the species or range of species being targeted in the project (as we outlined in the previous section on identifying the probable distributions of the biological populations under study). Regardless if one is focused on salamanders or elk, the physical areas surrounding the project locale must be considered in the planning process. After all, if you cannot control, or at least influence, a surrounding area and, at the same time, take into account effects from that surrounding area, your project's success will be substantially compromised.

Thus we think it is always optimal to start by considering the desired ecological condition (i.e., the outcome you desire) of the largest planning area, as well as the ecological processes that must be present or provided, as the initial step in developing a management plan. For example, it will be impossible to adequately plan for the distribution and condition of a particular species or group of species if an essential vegetation type and condition does not occur across the management units within the overall planning area. Also, if working within a riparian corridor where exotic (invasive) plants are an issue, seed sources from outside the planning area must be controlled, unless you can guarantee a constant removal program into the future.

Within the framework of the desired management outcome (ecological condition), the requirements of specific focal ecosystem elements and species must be planned and allocated across the management units. Morrison (2009) offered the following priority steps for planning.

- Determine and clearly articulate the desired ecological condition of the planning area.
- Develop a conceptual ecosystem model of the planning area, identifying major factors that will lead to the desired ecological condition. For example, consider conditions and dynamics in the watershed(s) or basin(s) surrounding and potentially impacting the planning area.
- Identify key ecological attributes (e.g., rainfall, fire) that drive the ecosystem model(s).
- Identify the suite of species characteristics of the desired condition.

- Identify target species (which can include a specific target species or a wide array of species across various taxa). Depending on the jurisdiction, there are often legal requirements to consider, such as special-status (e.g., threatened or endangered, migratory) species at the state and federal levels.
- Identify key constraints and stressors (e.g., availability of permanent water) that inhibit proper functioning of the pathways in the ecological model (thus preventing attainment of the desired condition).
- Identify constraints and stressors that can be alleviated through specific activities.
- Identify constraints and stressors that cannot be alleviated.
  - Reevaluate the desired condition and the associated conceptual model(s) in light of such constraints and stressors and modify them as necessary so they become more tractable.
  - Develop a plan to monitor the key ecological attributes at the overall project scale.
  - Determine a list of response variables, establish a quantitative goal, identify the likely time frame for attaining the desired condition, and identify any thresholds for additional action (see Step 3, below).

Even when the entity responsible for the management of land and animals is focused on a single species or small subset of species—e.g., because of legal requirements, or interest in hunted species—the presence and activity of other species must be considered for any project to be successful. "Success" is a project-specific term, dependent on the desired outcome of the practitioners, but it should be based on specific metrics (Step 3, below). For instance, as illustrated by Lopez et al. (2017:166), the goal of "improve quail habitat" can be approached in various ways, such as with a focus on vegetation management, and with or without predator control. Any responsible and rational plan would include an assessment of potential predators and their impact on quail, and it would clearly need to include a spatial scale, most likely outside the immediate project area (unless it was very large). In a similar example, actions to effectively recover populations of threatened western snowy plovers (*Charadrius nivosus nivosus*) along the West Coast of the United States need to consider the influence of predators, as well as the effectiveness and feasibility of predator control techniques, such as those evaluated in an expert panel by Marcot and Elbert (2015).

Step 2. *Delineate the project area.* As noted in Step 1, the management plan for each project area (e.g., management unit, such as a watershed) must be derived from the desired ecological condition for the overall planning area (e.g., basin). By assessing the current condition of each project area, planners can determine how to allocate target levels for each ecosystem component across the specified projects, based on the desired condition for the overall planning area. For example, a restoration plan for a defined project area may include an increase or decrease in the amount of a particular ecosystem element (e.g., goshawk habitat) depending on the conditions in the project area and what is currently available (or available for restoration) in other project areas. Additionally, because of the location of a project area, certain management actions (e.g., cowbird control) may be deemed infeasible (e.g., proximity to residential areas). Specific actions to pursue include the following.

- Describe the current vegetation and other environmental features.
- Identify the location and amount of special features (e.g., old-growth trees, riparian area, aspen).
- Identify constraints and stressors, both spatially and temporally.
- Compare the relevant special features with project-wide distribution, area, and condition.
- Develop species lists from the literature, available data, and expert knowledge.

  - Identify target species. Evaluate the site's potential to maintain each species and the constraints on maintenance.
- Develop a preliminary management plan.
  - Evaluate the efficacy of the preliminary management plan in relation to neighboring environmental conditions, constraints, and stressors, including those outside the project area.
  - Revise the management plan.
- Develop a monitoring plan at the project scale, to be integrated into Step 3, below.
  - Determine response variables, as well as establish quantitative goals, time frames for attainment, and thresholds for additional action (see Step 3).

Step 3. *Establish an adaptive management plan.* Based on the results of pre-project surveys and preliminary sampling, modifications of the preliminary project plans can be made to enhance the outcome(s) of management activity. Once the project has been initiated, monitoring should be incorporated, as it will then indicate if, and the degree to which, the desired conditions are being met. The system of planning, sampling, and modification has been formalized within the concept of adaptive management, or adaptive resource management (ARM). This is centered primarily on monitoring the effects of land-use activities on key resources and then employing those results as a basis for modifying activities to better achieve project goals (Walters 1986; Lancia et al. 1996). Although ARM was developed to learn about the overall ecological system in a geographic region of interest (e.g., a national forest), we are applying its principles to monitoring the effectiveness of any management planning process. The key here is a formal understanding of the major problems that could arise after a project has been initiated, and the development of a formal plan on how to deal with them, should the problem(s) come to fruition. In other words, this is not a trial-and-error approach in which potential problems are not fully taken into account.

Adaptive management is an iterative process whereby management practices are carefully planned, implemented, and monitored at predetermined intervals. If the outcomes are consistent with the predictions, then the project continues as planned. If the outcomes deviate from the predictions, however, management (1) can continue, (2) terminate, or (3) change, depending on project goals.

Morrison et al. (2006:406–412) summarized the application of adaptive management to habitat modification, explaining that an adaptive management approach entails (1) identifying areas of scientific uncertainty, (2) devising field management activities as real-world experiments to test that uncertainty, (3) learning from the outcome(s) of such experiments, and then (4) recrafting management guidelines, based on the knowledge gained. Management guidelines can include testable hypotheses, where monitoring and adaptive management studies equate to conducting the experiment, and a revision of the management guidelines equates to a reevaluation of the study results, in terms of testing the validity of the initial hypothesis (Morrison et al. 2006:407).

Trial-and-error approaches attempt to fix a problem after the implementation of a project and lack action plans formulated prior to its initiation. In contrast, adaptive management provides a structure whereby clear goals and objectives are established and monitored, and specific actions are planned at the outset of the project, in order to respond to deviations from the projected interim and final goals. This general procedure has been summarized in a 7-step process that includes feedback loops, which depend largely on monitoring results. The primary feedback loops in Figure 7.6 are between steps 5–6–7, 2–7, and 7–1. The 5–6–7 loop is the shortest and implies that management prescriptions are essentially working and need only slight adjustments. The primary obstacle in this loop is the time lag between the project's implementation and the point where the project has advanced sufficiently to monitor its effectiveness. Consequently, the loop is often severed and feedback is never provided. The second

loop, 2–7, indicates that monitoring was poorly designed, the wrong variables were measured, or monitoring was poorly implemented. The 7–1 feedback loop is when a decision must be made regarding the course of future management and monitoring activities. Here, if monitoring was done correctly, informed decisions can then be made for future management directions. If monitoring was done poorly, then another opportunity has been lost to provide a scientific basis for resource management.

For adaptive management to be effective, specific thresholds of key resources of interest must be established prior to the initiation of the project, which can include the stocking densities of plants, foliage cover, and the presence or a specific abundance of a target animal species. Violating such a threshold—either too high or too low—will then trigger a specific management action, assuming adequate resources are available to accomplish it. The results of ongoing monitoring should alert you to values that are approaching a threshold, so that preplanning can begin should the threshold be violated.

Thus the concept of thresholds and triggers are a central component of adaptive management and represent a predetermined level that, when violated, will lead to specific responses. Adaptive management plans center on those activities and outcomes that are central to the accomplishment of the project. For example, if a project is intended to promote successful breeding of an animal species in a degraded location, it is likely that vegetation goals must be met, as well as the potential suppression of predators, until the target level of reproduction is achieved. An adaptive plan would chart and then monitor the development of vegetation (e.g., density, height), and the outcome of breeding (e.g., nest success). Monitoring vegetation development annually would alert you if, say, the stocking density was falling below that required for the target animal species, and remedies would already have been anticipated and planned for (e.g., planting more trees, watering longer into the summer). Likewise, pre-project surveys would have alerted you that a potential predator would gain ac-

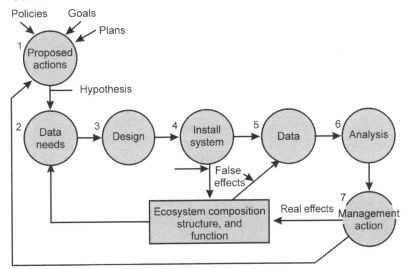

Social needs:

Policies   Goals

Plans

1 Proposed actions

Hypothesis

2 Data needs   3 Design   4 Install system   5 Data   6 Analysis

False effects

Ecosystem composition structure, and function   Real effects   7 Management action

Figure 7.6 A 7-step adaptive management system, showing the feedback loops that allow continual evaluation and adjustments of actions as the project progresses. See the chapter text for more detail.

cess to nests if water levels were too low. Again, you could have plans ready to be put into action as needed. The alternative—the way in which much of our management to date has been conducted—is to sit back and say, "What now?" Or, even if you had anticipated various problems and were monitoring for them, you would be left with the need to quickly find funds, materials, and personnel to help rescue the project. Your job—or future contract for similar work—might depend on it.

Thus adaptive management requires that, during the development of the management plan, potential actions be specified, so they could be undertaken if monitoring thresholds are triggered (see also Chapter 3). A failure to do negates many of the core benefits of adaptive management and results in a trial-and-error approach to resource management. Specific steps for developing an adaptive management approach include the following.

- Develop specific management actions if a threshold is triggered at either the overall project or subproject area scale.
  - o Implement additional management actions (e.g., more plantings or weedings, predator control).

  - o Modify the vegetation structure (e.g., change tree density).
  - o Revise treatment schedules (e.g., increase or decrease burning intervals or intensities).
- Revise the conceptual model or models as necessary, based on new information and changes in management actions and restoration activities.
  - o Revise thresholds and triggers, as indicated by the revisions in actions.
- Continue monitoring at appropriate spatial and temporal scales.

The concept of adaptive management also applies to research efforts in general. This is because, regardless of the initial care taken and the subsequent review of your study design, as well as your previous experience with similar work, field and lab studies seldom go as planned. As has been discussed in Chapter 3, methods are available for anticipating the necessary sample sizes most likely to meet your project goals. When studies last more than a few months (e.g., a single season of activity), however, it is wise to keep reevaluating your goals, the data being collected, and the ways in which you can improve your work. Unfortunately, in our experience few researchers adequately anticipate what might go wrong. Even if they do express various concerns, seldom do they have specific plans in place to quickly react to any contingencies. We realize that not all scenarios can be anticipated, and that not all management projects and research studies can be designed as statistical tests of every anticipated outcome. But there is no reason why major issues that could arise cannot be identified, and likely solutions preplanned. For example, because much of the work in animal ecology involves capturing and handling individuals, one can anticipate such issues as an inability to capture an adequate number of animals, a skewed sex ratio in those captured, equipment failure, and so on. It is best to preplan how you will handle such problems, especially in our often short-term and graduate student–based work.

Although researchers and managers are generally aware of the process of adaptive management, virtually no effort has been made during the development of monitoring plans to identify the biological populations that are the focus of the work, as well as the various other species that might interact with them. In other words, formulating a detailed adaptive monitoring plan does not necessarily improve the chances that a project will be successful (however success is defined) if the said project was crafted and implemented with little or no understanding of the biological populations involved. This is why, above, we first discussed assembling the species (and, hence, the geographic area needing to be encompassed by the study) that could influence the outcome of a project.

A Call for Ecological Reality

Ecologists are generally reluctant to identify the weaknesses of the design that led to the manuscript they are trying to publish. As we have posited throughout this book, ecologists in general have accepted the use of vague terminology and concepts, because that is "how we do things." We proceed in this manner because the other way—identifying the boundaries of what we are studying—is usually really difficult. We have argued that a main reason why many attempts to manage land and animals (e.g., habitat restoration) fail is because ecologists are not giving managers guidance that is based on what we will refer to as "ecological reality." Ecological jargon does not translate well (cannot be operationalized) into practical, on-the-ground management.

In this chapter we have outlined specific steps that most, if not all, ecological research can use as a framework for (1) identifying what we know and do not know about the system and the animal species under study, including the geographic extent(s) within which these processes and species are operating; (2) pinpointing the various filters and constraints that are operating within this appropriate geographic area; and (3) developing a detailed adaptive monitoring plan designed to enhance the success of management actions (which also feeds back into additional research, as needed). We are confident what we have presented represents a viable option. We say this because our concepts are based on a synthesis and adaptation of the many good ideas of numerous thoughtful individuals.

LITERATURE CITED

Anderson, A., J. E. Light, O. M. Takano, and M. L. Morrison. 2018. Population structure of the Townsend's big-eared bat (*Corynorhinus townsendii townsendii*) in California. Journal of Mammalogy 99:646–658.

Benedetti-Cecchi, L. 2001. Beyond BACI: Optimization of environmental sampling designs through monitoring and simulation. Ecological Applications 11(3):783–799.

Block, W. M. 2012. Analysis paralysis. Journal of Wildlife Management 76:875–876.

Bucklin, D. N., M. Basille, A. M. Benscoter, L. A. Brandt, F. J. Mazzotti, S. S. Romanach, C. Speroterra, and J. I. Watling. Comparing species distribution models constructed with different subsets of environmental predictors. Diversity and Distributions 21:23–35.

Cam, E., J. D. Nichols, J. R. Sauer, and J. E. Hines. 2002. On the estimation of species richness based on the accumulation of previously unrecorded species. Ecography 25:102–108.

Collier, B. A., J. E. Groce, M. L. Morrison, J. C. Newnam, A. J. Campomizzi, S. L. Farrell, H. A. Mathewson, R. T. Snelgrove, R. J. Carroll, and R. N. Wilkins. 2012. Predicting patch occupancy in fragmented landscapes at the rangewide scale for endangered species: An example of an American warbler. Diversity and Distributions 18:158–167.

Covert-Bratland, K. A., W. M. Block, and T. C. Theimer. 2006. Hairy woodpecker winter ecology in ponderosa pine forests representing different ages since wildfire. Journal of Wildlife Management 70(5):1379–1392.

Dennis, R. L. H., T. G. Shreeve, and H. V. Dyck. 2003. Towards a functional resource-based concept for habitat: A butterfly biology viewpoint. Oikos 102:416–426.

Duncan, D. H., and P. A. Vesk. 2013. Examining change over time in habitat attributes using Bayesian reinterpretation of categorical assessments. Ecological Applications 23:1277–1287.

Dunning, J. B., B. J. Danielson, and H. R. Pulliam. 1992. Ecological processes that affect populations in complex landscapes. Oikos 65:169–75.

Elith, J., and J. R. Leathwick. 2009. Species distribution models: Ecological explanation and prediction across space and time. Annual Review of Ecology, Evolution, and Systematics 40:677–697.

Elton, C. S. 1927. Animal ecology. Sidgwick & Jackson, London.

Fladder, S. L., and J. B. Callicott, eds. 1992. The river of the Mother of God and other essays by Aldo Leopold. University of Wisconsin Press, Madison.

Grinnell, J. 1917. The niche relationships of the California thrasher. Auk 34:427–433.

Guisan, A., and W. Thuiller. 2005. Predicting species distribution: Offering more than simple habitat models. Ecology Letters 8:993–1009.

Guthery F. S., and B. K. Strickland. 2015. Exploration and critique of habitat and habitat quality. Pp. 9–18 in M. L. Morrison and H. A. Mathewson, eds. Wildlife habitat conservation: Concepts, challenges, and solutions. Johns Hopkins University Press, Baltimore.

Hobbs, R. J., and D. A. Norton. 2004. Ecological filters, thresholds, and gradients in resistance to ecosystem assembly. Pp. 72–95 in V. M. Temperton, R. J. Hobbs, T. Nuttle, and S. Halle, eds. Assembly rules in restoration ecology: Bridging the gap between theory and practice. Island Press, Washington, DC.

Hunter, J. E., and F. L. Schmidt. 1990. Methods of meta-analysis. Sage, Newbury Park, CA.

Hurlbert, S. H. 1984. Pseudoreplication and the design of ecological field experiments. Ecological Monographs 54:187–211.

Hurlbert, S. H. 2009. The ancient black art and transdisciplinary extent of pseudoreplication. Journal of Comparative Psychology 123(4):434–443.

Hutchinson, G. E. 1957. Concluding remarks. Cold Springs Harbor Symposium on Quantitative Biology 22:415–422.

Johnson, C. J., and M. P. Gillingham 2005. An evaluation of mapped species distribution models used for conservation planning. Environmental Conservation 32:117–128.

Johnson, D. H. 2002. The importance of replication in wildlife research. Journal of Wildlife Management 66:919–932.

Lancia, R. A., C. E. Braun, M. W. Collopy, R. D. Dueser, J. G. Kie, C. J. Martinka, J. D. Nichols, T. D. Nudds, W. R. Porath, and N. G. Tilghman. 1996. ARM! for the future: Adaptive resource management in the wildlife profession. Wildlife Society Bulletin 24:436–442.

Leibold, M. A., M. Holyoak, N. Mouquet, P. Amarasekare, J. M. Chase, M. F. Hoopes, R. D. Holt, J. B. Shurin, R. Law, D. Tilman, M. Loreau, and A. Gonzalez. 2004. The metacommunity concept: A framework for multi-scale community ecology. Ecology Letters 7:601–613.

Lopez, R. R., I. D. Parker, and M. L. Morrison. 2017. Applied wildlife habitat management. Texas A&M University Press, College Station.

Lyet, A., W. Thuiller, M. Cheylan, and A. Besnard. 2013. Fine-scale regional distribution modelling of rare and threatened species: Bridging GIS tools and conservation in practice. Diversity and Distributions 19:651–663.

Marcot, B. G., and D. C. Elbert. 2015. Assessing management of raptor predation for western snowy plover recovery. General Technical Report PNW-GTR-910. US Department of Agriculture, Forest Service, Pacific Northwest Research Station, Portland, OR.

Mathewson, H. A., J. E. Groce, T. M. McFarland, M. L. Morrison, J. C. Newnam, R. T. Snelgrove, B. A. Collier, and R. N. Wilkins. 2012. Estimating breeding season abundance of golden-cheeked warblers in Texas. Journal of Wildlife Management 76:1117–1128.

Mathewson, H. A., and M. L. Morrison. 2015. The misunderstanding of habitat. Pp. 3–8 in M. L. Morrison and H. A. Mathewson, eds. Wildlife habitat conservation: Concepts, challenges, and solutions. Johns Hopkins University Press, Baltimore.

McFarland, T. M., J. A. Grzybowski, H. A. Mathewson, and M. L. Morrison. 2015. Presence-only species distribution models to predict suitability over a long-term study for a species with a growing population. Wildlife Society Bulletin 39:218–224.

McFarland, T. M., H. A. Mathewson, J. E. Groce, M. L. Morrison, and R. N. Wilkins. 2013. A range-wide survey of the endangered black-capped vireo in Texas. Southeastern Naturalist 12:41–60.

Mitchell, M. S., and R. A. Powell. 2002. Linking fitness landscapes with the behavior and distribution of animals. Pp. 93–124 in J. A. Bissonette and I. Storch, eds.

Landscape ecology and resource management: Linking theory with practice. Island Press, Washington, DC.

Morrison, M. L. 2009. Restoring wildlife: Ecological concepts and practical applications. Island Press, Washington, DC.

Morrison, M. L. 2012. The habitat sampling and analysis paradigm has limited value in animal conservation: A prequel. Journal of Wildlife Management 76:438–450.

Morrison, M. L., W. M. Block, M. D. Strickland, B. A. Collier, and M. J. Peterson. 2008. Wildlife study design, 2nd edition. Springer-Verlag, New York.

Morrison, M. L., B. A. Collier, H. A. Mathewson, J. E. Groce, and R. N. Wilkins. 2012. The prevailing paradigm as a hindrance to conservation. Wildlife Society Bulletin 36:408–414.

Morrison, M. L., B. G. Marcot, and R. W. Mannan. 2006. Wildlife-habitat relationships: Concepts and applications, 3rd edition. Island Press, Washington, DC.

Mouillot, D., O. Dumay, and J. A. Tomasini. 2007. Limiting similarity, niche filtering, and functional diversity in coastal lagoon fish communities. Estuarine, Coastal and Shelf Science 71:443–456.

Nuttle, T., R. J. Hobbs, V. M. Temperton, and S. Halle. 2004. Assembly rules and ecosystem restoration: Where to from here? Pp. 410–422 *in* V. M. Temperton, R. J. Hobbs, T. Nuttle, and S. Halle, eds. Assembly rules and restoration ecology: Bridging the gap between theory and practice. Island Press, Washington, DC.

Peña, D. 1997. Combining information in statistical modeling. American Statistician 51(4):326–332.

Pickett, S. T. A., J. Kolasa, and C. G. Jones. 1994. Ecological understanding. Academic Press, San Diego.

Pikesley, S. K., S. M. Maxwell, K. Pendoley, D. P. Costa, M. S. Coyne, A. Formia, B. J. Godley, W. Klein, J. Makanga-Bahouna, S. Maruca, S. Ngouessono, R. J. Parnell, E. Pemo-Makaya, and M. J. Witt. 2013. On the front line: Integrated habitat mapping for olive Ridley sea turtles in the southeast Atlantic. Diversity and Distributions 19:1518–1530.

Pulliam, H. R. 1988. Sources, sinks, and population regulation. American Naturalist 132:652–661.

Shirley, S. M., Z. Yang, R. A. Hutchinson, J. D. Alexander, K. McGarigal, and M. G. Betts. 2013. Species distribution modelling for the people: Unclassified Landsat TM imagery predicts bird occurrence at fine resolutions. Diversity and Distributions 19:855–866.

Smallwood, K. S. 2001. Linking habitat restoration to meaningful units of animal demography. Restoration Ecology 9:253–261.

Spake, R., and C. P. Doncaster. 2017. Use of meta-analysis in forest biodiversity research: Key challenges and considerations. Forest Ecology and Management 400:429–437.

Sundermann, A., S. Stoll, and P. Haase. 2011. River restoration success depends on the species pool of the immediate surroundings. Ecological Applications 21:1962–1971.

Temperton, V. M., and R. J. Hobbs. 2004. The search for assembly rules and its relevance to restoration ecology. Pp. 34–54 *in* V. M. Temperton, R. J. Hobbs, T. Nuttle, and S. Halle, eds. Assembly rules in restoration ecology: Bridging the gap between theory and practice. Island Press, Washington, DC.

Temperton, V. M., R. J. Hobbs, T. Nuttle, M. Fattorini, and S. Hallel. 2004. Introduction: Why assembly rules are important to the field of restoration ecology. Pp. 1–8 *in* V. M. Temperton, R. J. Hobbs, T. Nuttle, and S. Halle, eds. Assembly rules in restoration ecology: Bridging the gap between theory and practice. Island Press, Washington, DC.

Turner, M. G. 2005. Landscape ecology: What is the state of the science? Annual Review of Ecology, Evolution, and Systematics 36:319–344.

van Andel, J., and A. P. Grootjans. 2006. Concepts in restoration ecology. Pp. 16–28 *in* J. van Andel and J. Aronson, eds. Restoration ecology. Blackwell, Oxford.

Van Horne, B. 2002. Approaches to habitat modelling: The tensions between pattern and process and between specificity and generality. Pp. 63–72 *in* J. M. Scott, P. J. Heglund, M. L. Morrison, J. B. Haufler, M. G. Raphael, W. A. Wall, and F. B. Samson, eds. Predicting species occurrences: Issues of accuracy and scale. Island Press, Washington, DC.

Walters, C. 1986. Adaptive management of renewable resources. Macmillan, New York.

White, P. S., and A. Jentsch. 2004. Disturbance, succession, and community assembly in terrestrial plant communities. Pp. 342–366 *in* V. M. Temperton, R. J. Hobbs, T. Nuttle, and S. Halle, eds. Assembly rules in restoration ecology: Bridging the gap between theory and practice. Island Press, Washington, DC.

Wiens, J. A. 1989. The ecology of bird communities. Vol. 1, Foundations and patterns. Cambridge University Press, Cambridge.

Wiens, J. A. 2016. Ecological challenges and conservation conundrums: Essays and reflections for a changing world. John Wiley & Sons, West Sussex, UK.

Wiens, J. A., and B. T. Milne. 1989. Scaling of "landscapes" in landscape ecology, or, landscape ecology from a beetle's perspective. Landscape Ecology 3:87–96.

Young, M., and M. H. Carr. 2015. Application of species distribution models to explain and predict the distribution, abundance, and assemblage structure of nearshore temperate reef fishes. Diversity and Distributions 21:1428–1440.

Index

Page numbers followed by b, f, and t indicate boxes, figures, and tables, respectively.

abiotic filters, 97, 163–65, 164f
Adams, C. C., 19
Adams, D. C., 77, 86, 87
adaptation: vs. behavioral flexibility, 73; definition of, 71; and ecological continuity, 45; in ecotypes, 6, 7, 111; and evolutionary potential, 114, 114t, 116–25; factors contributing to, 72–73, 72f; individual variation in, 70–71; in metapopulations, 9; in neutral assembly, 87; in niches, 20; in population-level studies, 98; predicting future of, 73, 75; rates of, 71
adaptive co-management, 143–44
adaptive management, 140–47; challenges of using, 140, 144; complementary studies in, 144–47, 145f, 146f; establishing plans for, 175–78; feedback loops in, 176, 177f; tenets of, 140–44; thresholds and triggers in, 176–77
adaptive resource management (ARM), 175–76
administrative hierarchy, 35–36, 35f
aesthetic issues, in wildlife management, 3
agama lizards, toad-headed, 116, 116f
Agrawal, A. A., 55, 88
albatrosses, Laysan, 32
Alexander Archipelago, 38
algorithms, connectivity, 51–53
alligators, American, 132
Almpanidou, V., 117
alpha-diversity of vegetation structure, 45
Alström, P., 120
American alligators, 132
American bison, 132
American Ornithologists' Union, 5
American pikas, 115
ancient habitat concepts, 15–18, 16f, 24
Anderson, A., 169

animal ecology: individual organisms as foundation of, 2, 4; vs. natural history, 70; populations as foundation of, 8
animal rights, 3, 26
Anthony, C. D., 125
Anthropocene, 74–75
anthropogenic impacts. See human impacts
aquatic systems, landscapes of, 33
Aquinas, Thomas, 26
Arbogast, B. S., 120
Aristotle, 16
ARM (adaptive resource management), 175–76
Armitage, D. R., 143
Armstrong, D. P., 92–94
Ashley, M. V., 73
Asian elephants, 48f
Asian fairy bluebirds, 120
assemblages: definitions of, 74, 78; misuse of term, 6, 78, 78f
assembly, species, 86–88; development of concept of, 86–88; process of, 165, 166f; rules of, 86, 97, 161–63; in study design, 161–65
Aubry, K. B., 131
austral thrushes, 53

Bachman's sparrows, 41t
Baird's shrews, 132
Ball, M. C., 101–2
barred owls, 54
Barrett, K., 117
Bartelt, P. E., 119
bats, Townsend's big-eared, 169–70, 169f
bay-breasted warblers, 21f
Beebee, T. J. C., 124
behavioral flexibility, 73
behavioral research, 18

Beissinger, S. R., 56
Bell, D. M., 119
Berson, Henri, 27
big-eared bats, Townsend's, 169–70, 169f
bighorn sheep, desert, 70
binomial nomenclature, 17f
BIOCAP, 46
bioclimatic envelope, 117, 119
biocoenosis, 79
biodiversity, cryptic, 124
bioenergetics modeling, 119
biogeography: evolution and, 71; habitat connectivity in,
    50–51; island, 165; spatiotemporal scale of, 32, 37, 37f;
    species pools in, 162
biological organization: definition of, 4; as dimension of
    scale, 35f, 36; and human perspective, 3–5
biological populations. *See* population(s)
biomes, 37f, 38, 39f, 126
biophysical services, 138, 139t
biotic filters, 97, 163–65, 164f
biotic integrity, 126
birds: game management of, 24–25; migratory, metapopu-
    lations of, 91b–92b, 91f; standardization of names of, 5.
    See also specific species
bison, American, 132
bitterns, least, 130
black-capped vireos, 168
black sea turtles, 125
Block, W. M., 1, 12, 95
bluebirds, Asian fairy, 120
bobwhites, northern, 24
boundaries: of communities, 78–84, 80b–83b, 80f, 81f, 89;
    of ecosystems, 10–11; of habitats, 1–2, 84; of landscapes,
    32–33; of sampling areas, 95–97, 96f; of selective
    environment, 8–9, 84
boundary problem, viii; in community studies, 79–84,
    80b–83b, 80f, 81f; in habitat studies, 1–2
Bowen, B. W., 125
Brandon, R. N., 8
Brussard, P. F., 126
budworms, spruce, 57
buntings, painted, 123
Butler, R. S., 124
butterflies, Karner blue, 123

Callicott, J. B., 27
Cam, E., 165
canopy gaps, 56–57, 60
Cape May warblers, 21f
capture-recapture sampling, 22–23
caribou: ecotypes of, 7, 112; woodland, 102, 102f, 112,
    171b–72b, 172f

Carpenter, V., 18
Causal Interactionist Population Concept (CIPC), 9–10
character divergence, 72, 114t, 116, 119
Checklist of North American Birds, 5
CIPC (Causal Interactionist Population Concept), 9–10
Circuitscape, 52–53
classic definition of habitat, 156–57
classic study design, 167, 167f
clearcuts, 41t, 43, 44, 60, 61t
Clement, S., 75
Clements, F. E., 18
climate change: and disjunct populations, 121–23; as
    disturbances, 58; evolutionary perspective on, 74; and
    novel ecosystems, 86; spatiotemporal scale of, 37–38,
    39t; and species' range of tolerance, 117–19; in study
    design, 170
climate niche modeling, 117
closed systems: communities as, 79; existence of, 11;
    populations as, 99, 100
co-extinction, 115
Cohen, J. E., 84
cold spots, 134t
collection cabinets, 17
Collier, B. A., 168
Columbia River Basin: decline of native environments in,
    114; disjunct populations in, 120–21, 122f, 123f; key
    environmental correlates in, 128; range peripheries in,
    117, 118f; temporal scale of changes in, 38
co-management, adaptive, 143–44
Committee on the Status of Endangered Wildlife, 112
community/communities, 77–84; boundaries of, 78–84,
    80b–83b, 80f, 81f, 89; definitions of, 74, 77–79, 83b, 84;
    food webs in structure of, 84–86; gaps in study of, 86,
    88; local, 88–89; meta-, 88–89; misuse of term, 78–79,
    78f, 165
community ecology: boundary problem in, 79, 80b–83b,
    80f, 81f, 89; misuse of terms in, 78–79, 78f
community species pool, 162, 164f
competition: in CIPC, 9; in early habitat studies, 19; in
    ecosystem context, 74; exploitative, 84–85, 85f; and
    habitat heterogeneity, 46, 46f; interference, 85, 85f; in
    local extirpations, 115; in species assembly, 86–88, 161,
    163; as ultimate factor in survival, 2
complementary studies, in adaptive management, 144–47,
    146f
composite indices, 48
computers, in history of habitat studies, 21
connectivity, patch, 40, 41t, 47. *See also* habitat
    connectivity
conservation of species, political issues in, 3. *See also* habitat
    conservation
Copeland, J. P., 119

corridors, 53–54; definition of, 53; effectiveness of, 53–54; functions of, 53–54; least-cost, 51–52; in maintenance of evolutionary potential, 123; in metapopulation paradigm, 93; and patch shape, 48; in study design, 160, 162f, 174; vegetation, 40, 41t

Costanza, R., 138

Covington, W. W., 147

Crawford, H. S., 57

Crichton, Michael, 1

critical functional link species, 132–36, 134t, 136t

critical habitat, 25

Cronin, M. A., 7

Crumley, C. L., 147

cryptic biodiversity, 124

cryptic (sibling) species complexes, 114t, 116f, 120, 121f, 124–25

cultivated landscapes, 33–34

cultural issues, in wildlife management, 3

Darwin, Charles, 17–18

Darwin's finches, 58

decision modeling, 141

deer: Rocky Mountain mule, 44; white-tailed, 44

defoliating insects, 57

demes, 7–8

demonstration studies, in adaptive management, 144

Derryberry, E. P., 71

desert bighorn sheep, 70

design, study. See study design

Dewey, John, 27

Diamond, J. M., 86

directional selection, 72, 72f

disjunct populations, evolutionary potential of, 114t, 120–23, 122f, 123f

dispersal: in aquatic landscapes, 33; in communities, 83, 83f; and disturbances, 57; and habitat heterogeneity, 41t, 46–47, 53–54; in maintenance of species viability, 117, 120, 123; in metapopulations, 10; modeling of, 87, 124, 171f; of plants, 136t, 139t, 163; in population delineation, 107b; in population genetics, 101; in species assembly, 87–90, 91b–92b, 91f, 92, 163; in species diversity, 87; vector (agent), 74, 115, 125

dispersal-assembly models, 87

distinct population segment (DPS), 7, 112, 113f

disturbance(s), 54–61; adaptive management approaches to, 142, 143; in evolutionary perspective, 71, 73–75; factors affecting, 59–60; functional responses to, 135t; habitat tied to processes of, 31; management lessons from, 60–61; population linkages and, 123; spatiotemporal scale of, 35, 54; in species assembly, 87, 163; in SDMs, 171b, 171f; in study design, 170; and succession, 165; types of, 54–59, 55f

disturbance ecology, 43, 54

disturbance index, 48

Dobzhansky, T., 70

DPS (distinct population segment), 7, 112, 113f

drift. See genetic drift

drought, in Dust Bowl, 59

dusky-footed woodrats, 44

Dust Bowl, 59

Eckrich, C. A., 117

ecoclines, 40

ecological continuity, 44–45

ecological ethic, 27

ecological function, key. See key ecological function

ecologically functional groups, 128

ecological networks, 84

ecological processes, KEFs in, 130–31

ecological reality, 178

ecological relationships, history of study of, 19–20

ecological restoration. See restoration ecology, assembly rules in; restoration planning, steps for

ecological services. See ecosystem services

ecological units, lack of classification system for, 126

economics: in definition of wildlife, 3; in wildlife management, 3, 25–26, 35, 138

economic services, 138, 139t

ecosystem(s), 125–27; boundaries of, 10–11; challenges of identifying, 126; classification of, 126; definitions of, 10–11, 74; in evolutionary perspective, 74–75; functional redundancy in, 131, 131f; human impacts on, 74–75; landscapes in concept of, 12, 31–32; misuse of term, 126; novel, 75, 86; vs. species, as focus of wildlife management, 138; value of concept of, 126–27

ecosystem collapse, 126

ecosystem context, 74–75

ecosystem function, key. See key ecological function

ecosystem management, 125–31; assumptions in, 125; classification system for ecological units in, 126; components of, 74, 74f; definitions of, 126; KEFs in, 128–31; key environmental correlates in, 127–28, 130; science vs. goals/policies in, 126; spatiotemporal scale of issues in, 38–39, 39f, 127

ecosystem resilience, 126, 131

ecosystem services, 138, 139t

ecotones, 40

ecotypes, 111–12; adaptation in, 6, 7, 111; in biological organization, 35t, 36; in communities, 83; definitions of, 6, 7, 111; examples of, 7, 112; locally adapted, evolutionary potential of, 114t, 119–20, 123, 125; misuse of term, 111; in study design, 168; vs. subspecies, 7

edge contrast, 40, 41t

edge effects, 40, 41t

edge permeability, 40, 41t

effective population size, 92, 99, 102, 103b–5b

eigenvalue effective size, 99

Elbert, D. C., 175

elephants, Asian, 48f

Elith, J., 171b

elk, Rocky Mountain, 26f, 31, 43–44

El Niño, 58

Elton, C., 20, 159

Emery, N. C., 73

emigration, 46, 89–90

endangered species: concepts used in listing of, 7, 112, 113f; definition of, 112

Endangered Species Act (ESA), US: definition of endangered species in, 112; DPSs under, 7, 112, 113f; ESUs under, 7, 112, 125; evolutionary potential under, 125; habitat concepts in, 25; passage of, 25; SPR under, 112, 113f; terminology used in, 7, 112

endemic species: evolutionary potential of, 114t, 119; on islands, 119, 120, 141

endemic subspecies, evolutionary potential of, 114t, 119–20

endemism hotspots, 119

ensembles, 78, 78f

environment(s): classifying heterogeneity of, 8; populations in relation to, 8–10; selective, boundaries of, 8–9, 84

environmental correlates. See key environmental correlates

environmental gradients, in species abundance, 95, 95f

environmental movement, US, 25

ESA. See Endangered Species Act

Esler, D., 91, 91b–92b

ESU (Evolutionarily Significant Unit), 7, 112, 125

ethical issues: in emergence of novel ecosystems, 75; in environmental movement, 25; in habitat conservation, 26–27; in wildlife management, 3

ethnobiotic services, 138, 139t

ethology, 18

EU (Evolutionary Unit), 7

evolution: factors contributing to, 72–73, 72f; individual variation in, 70–71; by natural selection, 17–18; populations as unit of, 9, 70; predicting future of, 73, 75; rates of, 71; in selective environment, 8–9

evolutionarily enlightened management, 73

Evolutionarily Significant Unit (ESU), 7, 112, 125

evolutionary ecology, spatiotemporal scale of, 37, 37f

evolutionary perspective, 70–75; definition of, 71; ecosystem context in, 74–75; factors contributing to adaptive changes in, 72–73, 72f; management implications of, 72, 73–74

evolutionary potential, maintenance of, 112–25; cryptic species complexes in, 114t, 120; for disjunct populations, 114t, 120–23, 122f, 123f; for endemic species, 114t, 119; for endemic subspecies and locally adapted ecotypes, 114t, 119–20; full array of natural environments in, 114,

114t; habitat along edges of range of tolerance in, 114t, 117–19; habitat along peripheries of ranges in, 114t, 116–17, 116f, 118f; long-term viability of all species in, 114–16, 114t; polymorphic populations in, 114t, 124–25; population linkages in, 114t, 123–24

Evolutionary Unit (EU), 7

exotic species, 164

exploitative competition, 84–85, 85f

extinction(s), 115–16; co-, 115; of critical functional link species, 132; in definition of endangered species, 112; in ecological ethic, 27; functional, 115–16; global, 50, 115; habitat fragmentation and, 6, 46, 50; in metapopulations, 50, 90, 93; phyletic, 116; pseudo-, 116; quasi-, 115–16; rates of, 50, 111; in small populations, 70–71; spatiotemporal scale of, 32, 38; in study design, 165; in subpopulations, 91b, 103b

extinction vortex, 6, 71, 115

extirpations, local, 115

Fahrig, L., 46

Fattebert, J., 124

Fauth, J. E., 78–79

Federal Land Policy and Management Act, US, 25

feedback loops, in adaptive management, 176, 177f

finches, Darwin's, 58

fires: as disturbances, 57–60; suppression of, 59

fire salamanders, 117

Fischer, J., 3–4, 6, 12

fish: exotic species of, 164; species pools of, 163–64

Fish and Wildlife Service, US, 112

Fladder, S. L., 27

flying foxes, 55, 115–16

flying squirrels: common giant, 45; cryptic species complexes of, 120; Humboldt's, 120; northern, 43, 45, 120

food webs, 84–86, 85f

food web theory, 127

forests: canopy gaps in, 56–57, 60; clearcuts in, 41t, 43, 44, 60, 61t; complementary studies on, 144, 145f, 146f; disturbances in, 56–60, 61t; ecological continuity in, 44–45; functional redundancy in, 131, 131f; horizontal fragmentation in, 45

Forman, R. T. T., 33

Forsman, E. D., 43, 44

Foster, D. R., 55

foxes, kit, 117

fractal indices, 48–49

fragmentation: landscape, 12, 40; temporal, 44–45; vegetation, 40, 43–44. See also habitat fragmentation

fragmentation index, 48

FRAGSTATS, 46

Frankel, O. H., 72, 125

Franklin, J. F., 60

Fretwell, S. D., 50
function, key ecological. *See* key ecological function
functional analysis, 131–37
functional attenuation, 135t
functional bottlenecks, 134t
functional breadth, 134t
functional diversity, total, 133t
functional extinctions, 115–16
functional groups, ecologically, 128
functional homology, 133t
functional keystone species, 134t
functional linkages, 134t
functional profiles, 133t
functional redundancy, 131, 131f, 133t, 136, 137f
functional resilience, 135t
functional resistance, 135t
functional richness, 133t
functional shifting, 135t
functional specialists, 132, 134t
functional webs, 133t
fundamental niches, 159

Galetti, M., 116
game management, habitat concepts in, 19, 24–25
Ganzhorn, J. U., 17
gap dynamics, 43, 58
garter snakes, 112
genetic drift, 70, 72, 72f, 73, 83, 83f, 103b, 115
genetic gradients, 101–5
genetic neighborhoods, 101
genetic relatedness, in habitat connectivity models, 53
genetics, population, 99–106
geographic extent, 34, 35t
geographic information system (GIS), 36, 46, 51, 136
geopolitical species, 125
Giles, R. H., Jr., 48
Gillingham, M. P., 171b–72b
Gilpin, M., 50
GIS (geographic information system), 36, 46, 51, 136
global extinctions, 50, 115
Godron, M., 33
golden-cheeked warblers, 92, 93f, 97, 98f, 168
gophers, White Salmon pocket, 119
Grant, P. R., 58
graph theory, 46, 49
grasses: in Dust Bowl, 59; exotic, 33
great tits, 23
Greenberg, R., 119
green sea turtles, 125
grey nurse sharks, 115
Grinnell, J., 20, 159
Grootjans, A. P., 162, 165

groups, intraspecific, terminology applied to, 7
guilds: definition of, 78; misuse of term, 2, 5, 6, 78, 78f; predation within, 85, 85f; in species assembly, 161
Guisan, A., 171b, 172b
Guthery, F. S., 111, 156–57
Guthrie, W. K. C., 26

habitat, 15–27; ancient concepts of, 15–18, 16f, 24; boundaries of, 1–2, 84; components of, 31; critical, 25; definitions of, 15, 23, 31, 88, 156–57; micro- vs. macro-, 11–12, 41t, 56; misuse of term, 5, 11, 23, 156–57; niches in relation to, 20–21, 159–60; origins of concept of, 18–19; vs. populations, in study design, 155–56; populations in relation to, 70–75; resources defining, 2; societal use of concepts of, 24–25; species as function of, 131–32, 132f; as species-specific concept, 11, 40
habitat connectivity, 49–53; amount needed, 49–50; corridors in, 53–54; definitions of, 24, 49; functions of, 50–51; measures of, 49, 51–53; misuse of term, 24
habitat conservation: along edges of range of tolerance, 114t, 117–19; for endemic species, 114t, 119; ethical issues in, 26–27; along peripheries of ranges, 114t, 116–17, 116f, 118f
habitat corridors. *See* corridors
habitat domains, 84–85, 85f
habitat fragmentation, 40–49; categorization of, 40–43; as component of heterogeneity, 40, 41t; corridors in, 53–54; definitions of, 40, 41t; horizontal, 45; measures of, 46–49; misuse of term, 6, 40; positive effects of, 46; problems with concept of, 4, 6; as species-specific concept, 11, 40; temporal aspects of, 44–45; vs. vegetation fragmentation, 40, 43–44; vertical, 45–46; within-stand patchiness in, 45
habitat heterogeneity, 40–54; components of, 40, 41t; definitions of, 40; horizontal, 45; measures of, 8, 46–49, 47t; positive effects of, 46; problems with classification of, 8; temporal, 44–45; vertical, 45–46. *See also* habitat fragmentation; heterogeneity
habitat islands, 43, 50, 90, 91
habitat paradigm, 92–94
habitat patches, misuse of term, 90
habitat quality: classification of, 21–23; and habitat connectivity, 52–53
habitat selection, 11, 19–24
habitat studies, 15–27; ancient, 15–18, 24; vs. behavioral research, 18; boundary problem in, 1–2; computers in, 21; and ethics of habitat conservation, 26–27; habitat quality in, 21–23; habitat selection in, 19–24; hierarchical, 12, 19; origins of concept of habitat in, 18–19; societal use of, 24–25
habitat-wildlife relationships. *See* wildlife-habitat relationships
Hall, L. S., 11, 23–24, 78

Hallfors, M. H., 123
Hanski, I., 9, 50, 90, 93
hares, snowshoe, 130
Harmon, L. J., 71
Harms, K. E., 87
hawk eagles, mountain, 120
heterogeneity: of communities, 81b–82b; of environments, 8; of patches, 40. *See also* habitat heterogeneity
hierarchical definition of habitat, 156–57
hierarchical habitat studies, 12, 19
hierarchy: administrative, 35–36, 35f; vs. concentric circles, 157; of KEFs, 128, 129t–30t
Hilden, O., 20
hippopotamuses, 132
historic ecology, spatiotemporal scale of, 37, 37f
Hitchings, S. P., 124
Hobbs, R. J., 163
Hodges, K. E., 5, 6, 12
Holocene, 33, 37
home ranges: and community boundaries, 79; definition of, 21; habitat conservation along peripheries of, 114t, 116–17, 116f, 118f; with habitat fragmentation, 43, 44; history of study of, 21; in population delineation, 107b; in SDMs, 172b; in study design, 161, 165
horizontal fragmentation, 45
host-parasitoid food webs, 84
hot spots: of disjunct populations, 120–21; of endemism, 119; as functional linkages, 134t; in range peripheries, 117; of species richness, 36
Hull, D. L., 15
human(s): classification of, 17; perspective of, 3–5, 31–32
humane moralism, 26–27
human impacts: 114t, 135t; on disturbances, 55, 59; on ecosystems, 74–75; on endemic species on islands, 119; on extinctions, 115; on habitat heterogeneity, 40, 53
humanism, 26–27
Humboldt's flying squirrels, 120
hunting: ancient habitat concepts in, 15–16, 16f, 24; modern habitat concepts in, 18, 24–25; in wildlife management, 3
hurricanes, 50, 54–56, 55f
Huston, M. A., 46
Hutchinson, G. E., 20

Ibarguchi, G., 119
IBC (individual-based clustering) analysis, 101–2
ideal free distribution, 50
immigration, 46, 89, 90
imperiled function, 136t
inbreeding: effects of, 100, 102; models of, 103b–4b
inbreeding coefficient, 99
inbreeding depression, 72, 73, 123

inbreeding effective size, 99, 103b
Indian giant squirrels, 45
indicator variables, in adaptive management, 142–43
indices, of habitat heterogeneity, 46–49, 47t
individual-based clustering (IBC) analysis, 101–2
individual organisms: clarity of meaning of, 2; as foundation of animal ecology, 2, 4; variation of, in evolution, 70–71
insects, defoliating, 57
interference competition, 85, 85f
interspersion index of habitat heterogeneity, 47t, 48
intraguild predation, 85, 85f
intraspecific groups, terminology applied to, 7
invasive species, 57, 59, 164, 174
irrational definition of habitat, 156
islands: biogeography of, 165; boundaries of, 10; as closed systems, 100; cryptic species complexes on, 120, 121f; disturbances on, 55, 58; endemic species on, 119, 120, 141; habitat, 43, 50, 90, 91; models of, 4; polymorphic populations on, 124, 124f; species assembly on, 86

Jennings, D. T., 57
Johnson, C. J., 171b–72b
Johnson, D. H., 11
Jorgensen, E. E., 11–12

Kant, Immanuel, 26
Karl, S. A., 125
Karner blue butterflies, 123
KEF. *See* key ecological function
kernel estimators, 21
key ecological function (KEF), 128–37; definition of, 128; and ecological processes, 130–31; and ecosystem services, 138; functional redundancy in, 131, 131f, 133t, 136t, 137f; hierarchical classification of, 128, 129t–30t; identifying, 128–31; mapping of, 136, 137f; patterns of, 132–36, 133t–36t; in species-environment relationship model, 136–37
key environmental correlates, 127–28; definition of, 128; and ecological processes, 130; and ecosystem services, 138; identifying, 127–28; in species-environment relationship model, 136–37
keystone species, 116, 132, 134t
killer whales, 7, 112
Kirk, D. A., 18
kit foxes, 117
Klopfer, P. H., 17
Kublai Khan, 24

Lack, D., 19, 20
Lamas, T., 48
land-cover patterns vs. habitat, 11

land ethic, 26, 27

Landsat, 42

landscape(s), 32–39; boundaries of, 32–33; classification of, 33–34; definitions and use of term, 12, 32–33; dimensions of scale in, 34–36, 35f; in ecosystem concept, 12, 31–32; fragmentation of, 12, 40; heterogeneity in (see habitat heterogeneity); reasons for studying, 32

landscape ecology, 33–39; definitions of, 33, 160; development of, 160; dimensions of scale in, 34–36, 35f, 160; scales of disciplines used in, 36–38, 37f; in study design, 160–61

langurs, Nilgiri, 45

La Niña, 58

Lanyu scops owls, 141

Larch Mountain salamanders, 132, 132f

large infrequent disturbances (LIDs), 55, 55f

large scale, definition of, 34

lava lizards, 120, 121f

Lawton, R. O., 56–57

Lay, D. W., 19

Laysan albatrosses, 32

LCC (least-cost corridors), 51–52

LCP (least-cost path), 51–52

least bitterns, 130

least-cost corridors (LCC), 51–52

least-cost path (LCP), 51–52

Leathwick, J. R., 171b

legislation and regulation: environmental, 25; habitat concepts in, 25; in wildlife management, 3. See also specific laws

Leibold, M. A., 88–89

leopards, 124

Leopold, A., 11, 26, 46, 155; A Sand County Almanac, 27

Levins, R., 8, 9, 50

LIDAR (light detection and ranging), 46

LIDs (large infrequent disturbances), 55, 55f

Liebhold, A. W., 57

Liebig, Justus, 160

life histories, 24, 37, 37f, 100, 106, 113f

light detection and ranging (LIDAR), 46

Lindenmayer, D. B., 3–4, 6, 12

linkage disequilibrium, 99

Linnaeus, Carl, 17, 17f

Linnean system of nomenclature, 5

lion-tailed macaques, 45

little spiderhunters, 120

lizards: lava, 120, 121f; toad-headed agama, 116, 116f

lobsters, rock, 101

local communities, 88–89

local extirpations, 115

localities, 88, 89

local species pool, 162–65, 164f, 170

Locke, John, 26

loggerhead sea turtles, 117

Looijen, R. C., 79–83, 80b–83b

loop analysis, 49

Lopez, R. R., 174

Losos, J. B., 71, 120

Lucas, H. L., Jr., 50

Lyet, A., 173b

macaques, lion-tailed, 45

macrohabitat, 11–12, 41t, 56

Maes, J., 138

Mägi, M., 23

mainland-island metapopulation, 50

Maldonado, J. E., 119

managed landscapes, 33–34

Management Unit (MU), 7

Mannan, R. W., viii

maps: connectivity, 52–53; of KEFs, 136, 137f; species distribution, 171b–72b, 172f. See also zoning maps

map scale, 34, 35f, 36

Marco Polo, 24

Marcot, B. G., viii, 117, 120–21, 128, 131, 175

marine mammals, 36

marine organisms, 100, 129t

marine systems, 3, 33, 35

Markovian analysis, 49

Mathewson, H. A., 159

matrix, misuse of term, 90

MaxEnt, 52, 123

Maxwell, S. L., 115

May, R. M., 84

Mayden, R. L., 124

Mayr, E., 8

McCann, K. S., 84

McNay, R. S., 141

medieval habitat concepts, 17

megalopolis landscapes, 33–34

Merriam, C. H., 19

meta-analysis, 159

metacommunities, 88–89

metacommunity study design, 167, 167f

metapopulations, 89–94; definitions of, 7, 9–10; development of concept of, 9, 89–90; vs. habitat, paradigms of, 92–94; habitat connectivity and, 50; of migratory species, 91–92, 91b–92b, 91f; patch terminology and, 90; spatiotemporal scale of stability in, 37, 37f; structure of, 92, 124

Mexican spotted owls, 19f

microbursts, 56

microevolution, 73

microhabitat, 11–12, 41t, 56

microserules, 56

microsites, 88, 89

migration: in habitat connectivity models, 53; habitat fragmentation and, 46; history of study of, 16; in idealized populations, 99–100; in metapopulations, 10, 89–90, 91–92, 91b–92b, 91f, 100; in subpopulations, 103b

Migratory Bird Hunting Stamp Act, US, 25

migratory species, metapopulations of, 91–92, 91b–92b, 91f

Millstein, R. L., 8–9, 11

Milne, B. T., 160

Mitchell, M. S., 160

MODIS, 42

molecular genetics, 101–2

Montoya, J. M., 84

moralism, humane, 26–27

More, Thomas, 26

Moritz, C., 112

Morris, D. W., 11

Morrison, M. L., viii, 1, 8, 12, 78, 95, 155, 159, 163, 173–76

Moser, D., 48

Mouillot, D., 163–64

mountain hawk eagles, 120

mountains, endemic species on, 119

Mt. Graham red squirrels, 141

MU (Management Unit), 7

mule deer, Rocky Mountain, 44

mutualistic food webs, 84

Myers, J. A., 87

National Environmental Policy Act, US, 25

National Forest Management Act, US, 25

natural, definitions of, 74, 75

natural history: vs. animal ecology, 70; habitat in study of, 16–17, 127, 156; history of study of, 16–17

natural landscapes, 33–34

natural resource services, 138, 139t

natural selection, evolution by, 17–18

networks: community, 88–89, 167; ecological, 84; modeling of, 52, 169f, 172b; researchers, 144; road, 52

neutral assembly, 87

neutral selection, 72, 72f, 73

niches: definitions and use of term, 20–21, 159; development of concept of, 20–21, 21f, 159; fundamental vs. realized, 159; in species assembly, 87; in study design, 159–60

Nilgiri langurs, 45

Noon, B. R., 44

northern bobwhites, 24

northern flying squirrels, 43, 45, 120

northern spotted owls, 6, 43, 44, 115

Northwest Power and Conservation Council, 128

Norton, D. A., 163

novel ecosystems, 75, 86

observational studies, in adaptive management, 144

occupancy modeling, 22

Odum, E. P., 20

Ohman, K., 48

old-growth forests, 6, 41t, 44–45, 59, 126, 175

Oliver, F. W., 18

olive Ridley turtles, 173b

open systems: boundaries of, 10–11; communities as, 79

Orrock, J. L., 48

Orsini's vipers, 173b

owls: barred, 54; Lanyu scops, 141. *See also* spotted owls

Pacific salmon, 99, 100, 112

painted buntings, 123

paleoecology, 35t, 37–39, 37f, 147

parrots, Puerto Rican, 56

Parsons, W. F., 57

patch: definitions of, 88; misuse of term, 90

patch connectivity, 40, 41t, 47

patch dynamics, 40, 41t

patch heterogeneity vs. habitat heterogeneity, 40

patchiness, within-stand, 45

patchiness index of habitat heterogeneity, 47t, 48, 48f

patch isolation, 40, 41t

patch mosaic paradigm, 40

patch shape index of habitat heterogeneity, 47–48, 47t

patch type richness and diversity, 40, 41t

patchy populations, 10

Pellet, J., 90

percolation threshold, 49

performance standards, 140–41

perspective: and biological organization, 3–5; definition of, 4; and landscape, 31–32

pests, 26–27

Peters, R. H., 126

pheasants, 24

phyletic extinctions, 116

phylogeography, 37, 147

phytogeography, 37, 37f

Pierson, E. D., 55

pikas, American, 115

Pikesly, S. K., 173b

pines, ponderosa, 59, 101, 105, 117

Pittman-Robertson Federal Aid in Wildlife Restoration Act, US, 25

pixels: in connectivity analysis, 51–52; in fragmentation analysis, 42–43; spatial resolution of, 34, 35t

plain-backed thrushes, 120

planning areas, delineation of, 173–75

Pleistocene: habitat concepts in, 15–16, 16f, 18; habitat connectivity in, 50–51

plovers, western snowy, 175

pocket gophers, White Salmon, 119

political issues: in adaptive management, 140–41; in ecosystem management, 126; in wildlife management, 3

polymorphic populations, evolutionary potential of, 114t, 124–25, 124f

ponderosa pines, 59, 101, 105, 117

population(s), 6–10; approaches to delineating, 95–106, 107b; in CIPC, 9–10; vs. communities, as focus of studies, 85; components of, 7; definitions of, 6–10, 74, 88, 98–99; in DPSs, 112; environment in relation to, 8–10; as genetic construct, 6, 99–106; vs. habitat, in study design, 155–56; habitat in relation to, 70–75; patchy, 10; polymorphic, 114t, 124–25; terminology of, 6–10; as unit of evolution, 9, 70. *See also* metapopulations; subpopulations

population-based study design, 167–68, 167f

population ecology: habitat vs. metapopulation paradigms in, 92–94; recommendations for improving research in, 94; spatiotemporal scale of, 37

population genetics, 99–106

population linkages, in maintenance of evolutionary potential, 114t, 123–24

population size: effective, 92, 99, 102, 103b–5b; genetic, 99; small, and extinctions, 70–71

population structure: gaps in knowledge of, 90, 94; of metapopulations, 92, 124; in study design, 168; of subpopulations, 90, 92

population viability: maintaining, 114–16, 114t; spatiotemporal scale of, 37, 37f

porosity index of habitat heterogeneity, 47t, 48

Post, D. M., 10

potholes meadow voles, 119

Powell, R. A., 160

predation: intraguild, 85, 85f; predator-prey interactions, 84–85, 85f; in study design, 174–75

Priestly, J. H., 18

project areas, delineation of, 175

proximate factors, 2

pseudoextinction, 116

pseudoreplication, 167

Puerto Rican parrots, 56

Putz, F. E., 56–57

quasi-extinctions, 115–16

random mating, in populations, 99, 100–101

ranges. *See* home ranges

Ray, John, 17

realized niches, 159

red-backed salamanders, 125

red-cockaded woodpeckers, 56

red squirrels, Mt. Graham, 141

red tree voles, 45

Regan, T., 27

regional species pool, 162–65, 164f, 170

regions, definitions of, 89

Renaissance habitat concepts, 17

reproductive interactions: in definition of populations, 9; random, 99, 100–101

resilience: in behavioral flexibility, 73; disturbances and, 59; ecosystem, 126, 131; of endangered species, 112, 113f; functional, 135t

resistance: functional, 135t; in functional redundancy, 133t; in habitat connectivity, 51, 52; wind, 60

resources, habitat defined by, 2

resource selection, 21–23

restoration ecology, assembly rules in, 163

restoration planning, steps for, 173–76

retrospective studies, in adaptive management, 144–47

Rice, K. J., 73

Ridley turtles, olive, 173b

ringneck snakes, 121, 123f

risk analysis, in adaptive management, 141, 143

risk management, in adaptive management, 140–41, 143

rock lobsters, 101

Rocky Mountain elk, 26f, 31, 43–44

Rocky Mountain mule deer, 44

Roman habitat concepts, 16

Rooney, N., 84

sage-grouse, western, 130

Sakia, H. F., 44

salamanders, 78f, 174; fire, 117; Larch Mountain, 132, 132f; red-backed, 125

salmon, 99, 100, 112, 125, 164

sampling, capture-recapture, 22–23

sampling areas: guidelines for choosing, 168–70; spatial extent of, 13, 95–97, 96f

Sand County Almanac, A (Leopold), 27

scale, spatiotemporal: dimensions of, 34–36, 35t; of disciplines used in landscape ecology, 36–38, 37f; of disturbances, 35, 54; of ecosystem management issues, 38–39, 39f, 127; in landscape ecology, 34–38, 160; misuse of term, 34

Schaus, W., 18

Schlaepfer, D. R., 119

Schmidt, K. A., 45

Schmitz, O. J., 84–85

Schoener, T. W., 20

scientific uncertainty, in adaptive management, 140, 143

SDM (species distribution model), 170, 171b-73b, 171f, 172f

sea turtles: black, 125; green, 125; loggerhead, 117; olive Ridley, 173b

selection, 2, 8; in communities, 83, 83f; directional, 72, 72f; habitat, 11, 19–24; natural, 17–18; neutral, 72, 72f, 73; resource, 21–23; in species assembly, 87–88; in species diversity, 87
selective environment: boundaries of, 8–9, 84; definition of, 8
semi-natural landscapes, 34
shags, Stewart Island, 124–25, 124f
shape index of habitat heterogeneity, 47–48, 47t
sharks, grey nurse, 115
sharptail snakes, 132
sheep, bighorn, 70, 112
Shipley, J. R., 123
Shirley, S. M., 173b
shortleaf pine–bluestem forests, 144, 145f, 146f
shortwings, white-bellied, 119
shrews, Baird's, 132
sibling (cryptic) species complexes, 114t, 116f, 120, 121f, 124–25
significant portion of the range (SPR), 112, 113f
Sims, K. E., 140
Singer, P., 27
sister (cryptic) species complexes, 114t, 116f, 120, 121f, 124–25
small populations, extinctions in, 70–71
Smallwood, K. S., 95, 158
snakes: garter, 112; ringneck, 121, 123f; sharptail, 132
snowshoe hares, 130
societal use of habitat concepts, 24–25
soil subsystems: ecological processes in, 130–31; KEFs in, 136, 137f; spatiotemporal scale of genesis and renewal in, 37–38, 37f
Sole, R. V., 84
Soulé, M. E., 72, 125
source-sink dynamics, 95
sparrows: Bachman's, 41t; white-crowned, 71
spatial resolution, 34, 35f
spatiotemporal scale. *See* scale, spatiotemporal
specialists, functional, 132
speciation: in communities, 83, 83f, 84, 87–88; cryptic biodiversity and, 124; habitat connectivity and, 50; means of, 10, 74; spatiotemporal scale of, 32; in species assembly, 87–88; in species diversity, 87–88; sub-, 88, 114t, 116
species: vs. ecosystems, as focus of wildlife management, 138; as function of habitat, 131–32, 132f; KEFs of, 128–31, 129t–30t; political issues in conservation of, 3
species abundance, environmental gradients in, 95, 95f
species assembly. *See* assembly, species
Species at Risk Act, Canadian, 112
species distribution model (SDM), 170, 171b–73b, 171f, 172f
species diversity: food webs and, 84; habitat heterogeneity and, 46, 46f; processes affecting, 87–88
species-environment relationship model, 136–37

species pools, 162–65, 162f, 164f, 170
spiderhunters, little, 120
Spiller, D. A., 55
spotted owls: Mexican, 19f; northern, 6, 43, 44, 115; political issues in conservation of, 3; population delimitation for, 107b
SPR (significant portion of the range), 112, 113f
spruce budworms, 57
squirrels: Indian giant, 45; Mt. Graham red, 141. *See also* flying squirrels
Standish, R. J., 75
state-space models, 53
statistics, technological advances in, 21
St. Clair, C. C., 47
Steen, D. A., 117
stepping-stone model, 101
Stewart Island shags, 124–25, 124f
Stoddard, H. L., 24
Stone, C. D., 27
Strickland, B. K., 156–57
Stroud, J. T., 120
study design, 155–78; definitions of habitat in, 156–57; framework for, 161–65; implementation of, 166–78; landscape ecology in, 160–61; major approaches to, 166–68, 167f; niche concept in, 159–60; populations vs. habitats in, 155–56; species assembly in, 161–65; species pools in, 162–65, 170; temporal and spatial considerations in, 156–59
subpopulations: definition and use of term, 2, 4, 7, 10; habitat connectivity and, 50; interaction between vs. within, 10, 124, 166; structure of, 90–94, 91b–92b, 103b, 105, 107b
subspecies: definition and use of term, 6, 7; vs. ecotypes, 7; endemic, evolutionary potential of, 114t, 119–20
suburban landscapes, 33–34
succession: in adaptive management, 145f, 146f; development of concept of, 19, 165; disturbances and, 54, 56, 58, 60; in habitat heterogeneity, 41t, 45, 49; key environmental correlates and, 128; origins of term, 18; spatiotemporal scale of, 37, 37f; species assembly and, 87, 161, 163, 166f; in study design, 165, 170
Sundermann, A., 163
survival interactions, in definition of populations, 9
Svärdson, G., 19, 20

taxonomy, 3, 114, 117, 170; in definition of communities, 79; in evolutionary perspective, 71–74; history of, 17, 17f; terminology of, 7
technology, in history of habitat studies, 21
Teilhard de Chardin, Pierre, 27
Temperton, V. M., 163
temporal fragmentation, 44–45
temporal scale. *See* scale, spatiotemporal

terminology: of ecosystems, 10–11; of habitat, 11–12; importance of, 5–6; of landscapes, 12; of populations, 6–10; standardization of, 5; vague and missing definitions of, vii, 2, 24
terrestrial landscapes, classification of, 33–34
terrestrial vertebrates, in concept of wildlife, viii, 2
Thomas Aquinas, 26
Thompson, J. N., 45
Thorpe, W. H., 19
thrushes: austral, 53; plain-backed, 120
Thuiller, W., 171b, 172b
time frames, evolutionary, 71, 73–74
time period, 35, 35f
tits, great, 23
toad-headed agama lizards, 116, 116f
toads, western, 119
tolerance, habitat conservation along edges of range of, 114t, 117–19
total functional diversity, 133t
Townsend's big-eared bats, 169–70, 169f
Tracy, C. R., 126
Trani, M. K., 48
treefalls, 58
tree voles, red, 45
trout, 164
turkey vultures, 132
Turner, M. G., 160
turtles. See sea turtles
Type I disturbances, 54–56, 55f
Type II disturbances, 55f, 56–58
Type III disturbances, 55f, 58
Type IV disturbances, 55f, 58–59

ultimate factors, 2
uncertainty: in adaptive management, 140, 143, 176; in definition of habitat, 18; in ecological conditions, 32
unsuitable habitat, 41t, 90
urban landscapes, 33–34
urn models, 100–101

van Andel, J., 79–83, 80b–83b, 162, 165
Van Balen, J. H., 23
Van Horne, B., 23, 95
variance effective size, 99
vegetation, in concept of habitat, 19, 19f
vegetation corridors, as component of heterogeneity, 40, 41t
vegetation fragmentation, 40, 43–44
vegetation structure, alpha-diversity of, 45
Vellend, M., 83, 87–88
Vergara, P. M., 53
vertebrates, terrestrial, in concept of wildlife, viii, 2
vertical fragmentation, 45–46

vipers, Orsini's, 173b
vireos, black-capped, 168
vital rates, 93–94
Vögeli, M., 93
voles: potholes meadow, 119; red tree, 45
vultures, 132

Wade, A. A., 24
Waples, R. S., 112
warblers: bay-breasted, 21f; Cape May, 21f; golden-cheeked, 92, 93f, 97, 98f, 168; yellow-rumped, 21f
Webb, S. L., 57
Weiher, E., 86, 87
western sage-grouse, 130
western snowy plovers, 175
western toads, 119
whales, killer, 7, 112
white-bellied shortwings, 119
white-crowned sparrows, 71
White Salmon pocket gophers, 119
white-tailed deer, 44
Wiens, J. A., 31, 77, 78, 85–86, 160, 165
wildlife: connotations of term, viii, 2; definition and use of term, 2–3; value of, 25–27
wildlife-habitat relationships: history of studies of, viii, 1–2, 155; KEFs in, 131–36; landscape approach to, 32
Wildlife-Habitat Relationships (Morrison, Marcot, & Mannan), viii
wildlife management: adaptive approach to, 140–47; disturbances in, 60–61; functional analysis in, 131–37; physical boundaries in, 10; species included in, 2, 3; species vs. ecosystems as focus of, 138
Wildlife Society, The, 125
Williams, D. W., 57
wind events, 56–57
within-stand patchiness, 45
wolverines, 32, 119
Woodhouse, S. W., 18
woodland caribou, 102, 102f, 112, 171b–72b, 172f
woodpeckers, red-cockaded, 56
woodrats, dusky-footed, 44
Wright, S., 101
Wüest, R. O., 117–19
Wunderle, J. M., Jr., 56

Yapp, R. H., 18
yellow-rumped warblers, 21f

zoning maps: of scale of disciplines used in landscape ecology, 36–38, 37f; of scale of ecosystem management issues, 38–39, 39f
zoogeography, spatiotemporal scale of, 37, 37f